BEAUTIFUL
Ferncliff

BEAUTIFUL
Ferncliff

*Springfield, Ohio's
Historic Cemetery & Arboretum*

by Paul W. Schanher, III
and Anne E. Benston

ORANGE FRAZER PRESS
Wilmington, Ohio

Copyright © 2008 The Springfield Cemetery Association & The Harry M. & Violet Turner Charitable Trust
ISBN: 978-1933197-531

No part of this publication may be reproduced in any material form (including photocopying or storing in any medium by electronic means and whether or not transiently or incidentally to some other use of this publication) without the written permission of the copyright holder except in accordance with the provisions of the Copyright, Designs and Patents Act 1988.

Published for the Turner Foundation by:

Orange Frazer Press
P.O. Box 214
Wilmington OH 45177
1.800.852.9332
www.orangefrazer.com

The Turner Foundation
4 West Main Street
Suite 800
Springfield, Ohio 45502
www.hmtunerfoundation.org
937.325.1300

Project director for The Turner Foundation Tamara K. Dallenbach
Book and cover design by Chad DeBoard

Library of Congress Cataloging-in-Publication Data

Schanher, Paul W. (Paul Welstead), 1946-
  Beautiful Ferncliff : Springfield, Ohio's historic cemetery &
arboretum / by Paul W. Schanher III and Anne E. Benston.
     p. cm.
  Includes bibliographical references and index.
  ISBN 978-1-933197-53-1 (alk. paper)
  1. Ferncliff Cemetery (Springfield, Ohio) 2. Springfield
(Ohio)--History. 3. Springfield (Ohio)--Biography. I. Benston, Anne E.
(Anne Elizabeth), 1923- II. Title.
  F499.S7S34 2008
  363.7'50977149--dc22
                        2008020552

Printed in China

DEDICATION

To Stanley Spitler with much gratitude.

ACKNOWLEDGEMENTS

My deepest appreciation and gratitude goes to the many who have provided assistance and encouragement in this endeavor.

First and foremost I am greatly indebted to Walt Kendig who was instrumental in my participation with the Ferncliff Cemetery exploratory committee. The committee was formed to "bring to life" the history and great beauty of the cemetery. That effort has now evolved into the Annual "Ferncliff in the Fall "event. The seed was planted at that time for a publication of the cemetery history, and a history of many of the long forgotten souls who have given labor and love for the betterment of the community and have now vanished from this earth.

Connie Chappell and Dan Dewine have enlightened me on the history of the early police chiefs dating from William Stewart, Springfield's first in 1867, James Fleming in 1871 and James Walker in 1883. I am humbled by their generosity.

A warm and heartfelt thank you to Tom Stafford, Community News Staff Writer for the Springfield News-Sun for his historical writings, enabling me to gain more background for many of the biographies. His weekly writings are a treasure to all of Clark County. Thank you, my friend.

It is with great pride that I acknowledge the Clark County Historical Society Archives under the leadership of Virginia Weygandt, director of collections, Kasey Eichensehr, curator and Natalie Fritz, curatorial assistant. They have worked non-stop scanning the material needed for the book with great patience and dignity and always a smile.

How will I ever be able to show my gratitude and love to Tamara Dallenbach, of The Turner Foundation for her belief in the proposed publication by my co-author and me? With her knowledgeable background in history she saw the possibility of a book and approached the executive director of the Turner Foundation, John Landess, for approval. With all the wonderful work the Turner Foundation has committed to the preservation of local history, one more project was approved with eternal gratitude.

From day one the Ferncliff Cemetery Board were always gracious and accepting in the almost daily browsing and researching of their records even on their busiest days. The board was also in favor of the publication and approved a commitment for funding with the Turner Foundation.

It was truly a magical day when my co-author, Dr. Paul "Ski" Schanher and I recognized we were on the same path in our aspiration to bring the history of Ferncliff Cemetery to the public. From the onset to the final chapter it was a joy to be able to work with someone who was so intensely dedicated to this project. It has been an honor to share this book with him.

Last, but far from least, I wish to thank my family for their strong support. They gave me encouragement to carry on when my days were long. I am so grateful to my husband, Chuck, for his limitless faith and patience, for being there and lifting the load of worry on those intense days when I felt I could not meet my imaginary deadline. Thank you, one and all.

Anne Benston

A well-known Lincoln historian once said, "It hardly seems fair that the intellectual debts I have piled up can be paid in so free and easy currency as mere acknowledgments." But so be it, as these next few sentences will serve as a public affirmation of those debts.

First and foremost, the principle debt to be paid is that to our editor Kim Byrum Skinner. Without her constant encouragement and gentle prodding, this manuscript could not have come to pass. Once handed portions of our writings over a several year period of time, Kim very precisely and expertly wove our words into a free-flowing work that actually makes some sense. Her professional manner in which she arranged what was handed to her brought the entire project into a readable manuscript that makes us sound better than we really are. It has always been our premise that to surround yourself with good people makes you better. Kim is good people. Our first debt is owed to her.

The superintendent of Ferncliff Cemetery, Stanley Spitler, opened the doors to the archives deep beneath the floors of the administration building so that we could search the original, handwritten pages of the minutes of the initial organizational meeting that convened with the first members of the Springfield Cemetery Association in June of 1863. His faithful assistant Scott Brown provided much insight into specific locations of several obscure burial sites, specifically that of the reported first burial in the cemetery, that of George Clemens. To Beth Anderson, Ferncliff's bookkeeper, we owe a major debt of gratitude for her kindness in assisting us in any way that she could. The support of the Springfield Cemetery Association and its Board of Trustees with their overwhelming approval and support of this project cannot possibly go without mention. The friendliness of those who carefully manage the grounds in all kinds of weather certainly made our research time there very pleasant.

The debt list must also include Pastor Eric Mounts of the Southgate Baptist Church, longtime friend Dan Powell of Grace Evangelical Lutheran Church, former racquetball opponent Grant Edwards of Fellowship Christian Church, and Retired Pastor Marv Wiseman for generously sharing their time and wisdom regarding the subject at hand.

Traveling to Ft. Wayne, Indiana to spend time with dear friend Tom Hopewell, principal of a major high school in that community, brought special insights as we discussed the beauty of the landscape architecture of the cemetery as well as eternal things. The time spent with Jerry Feinstein of Skokie, Illinois was precious. As we sat in a dormitory room at Gettysburg College during the Civil War Institute held there annually the last week of June, the tape recorder was challenged to keep up with his exuberance as he discussed his work in the Civil War section of a well-known cemetery in suburban Chicago. Dr. Rodney Wyse, retired professor in the business department of Central State University, spent a thought-provoking time with me in Bob Evans Restaurant over breakfast on a Friday morning as we discussed the issues at hand, not only about personal cemetery experiences, but also more importantly about life and striving to make a difference whether it be at home, work, or play. Much was gleaned form this wise sage of experience.

The debt list must also include Pam Corle-Bennett who wrote the arboretum chapter. Pam is the State Master Gardener Volunteer Coordinator, Horticulture Educator for The Ohio State University Extension, Clark County. Her extensive knowledge of all things botanical has leant an immense contribution to the book by way of her informative chapter on one of Ferncliff's greatest assets – the abundance and variety of trees that grace its hallowed grounds.

An enormous debt of gratitude is due to Douglas Keister, gifted photographer and author of *Stories in Stone: A Field Guide to Cemetery Symbolism and Iconography*. The examination in this book of symbolism at Ferncliff is patterned, with his gracious approval, on that seminal text.

George Berkhofer, author of a most recently

published book entitled *No Place Like Home*, spent a beautiful autumn afternoon with me traveling around the hills and dales of Ferncliff Cemetery with the tape recorder endlessly positioned on "play" as he described the architectural designs of each magnificently built mausoleum interspersed throughout the older sections of the cemetery. A particular word of thanks goes to son Seth Schanher for rewriting the chapter of the book concentrating on the symbolism carved into particular monuments commemorating certain aspects of the life of the individual laid to rest. To Virginia Weygandt, director of collections of the Clark County Historical Society, Kasey Eichensehr, curator, and Natalie Fritz, curatorial assistant, a gracious word of thanks for their tireless work in gathering pictures and encouraging us along the way.

Words cannot express our gratitude to The Turner Foundation, with John Landess as its executive director, for their undying support throughout this four-year project. Without what they do, this publication would not be possible. Our project director and primary liaison, Tamara Dallenbach, public historian to the Turner Foundation, proved to be a most invaluable person as she took care of the remotest detail and served as our main cheerleader. Without her experience and expertise, this publication would not be possible. Kevin Rose, an up and coming historian with the Foundation, gave much expert advice and provided several picture ideas that enhanced the overall integrity of this book.

I remember my first visit to Ferncliff as an interested bystander, after an extensive study of Springfield's history and the men and women who made it great. I was riding through Ferncliff Cemetery with Stan Spitler as he drove the minivan and with co-author and mentor Anne Benston at his side pointing out the final resting places of those who I felt as if I knew personally. What a wonderful spring day that was! From that experience Anne, noticing my excitement as Stan navigated the winding, hilly roads with its twists and turns, said, "Let's write a book."

With that, we began a four-year long journey together that eventually not only produced this book, but also solidified our friendship for all of eternity. To her I have no words further to say but "Thank You!!!"

Finally, no acknowledgment would be complete without thanking my family for their support and perseverance during this time. To my wife Cheryl, our daughters Stacy Berge and Stephanie Girth, and our son Seth, again I have no other words to say but "Thank You!!!"

Paul W. Schanher, III

The cemetery is a memorial and a record. It is not a mere field in which the dead are stowed away unknown; it is a touching and beautiful history, written in family burial plots, in mounded graves, in sculptured and inscribed monuments. It tells the story of the past, not of its institutions, or its wars or its ideas, but of its individual lives—of its men and women and children, and of its household. It is silent, but eloquent; it is common, but it is unique. We find no such history elsewhere; there are no records in all the wide world in which we can discover so much that is suggestive, so much that is pathetic and impressive.

—Joseph Anderson

TABLE OF CONTENTS

1. Rural Cemeteries — 1

2. Pioneer Spirit — 7

3. A Civic Duty Fulfilled — 16

4. Home is Where The Heart is — 25

5. Springfield Mourns — 31

6. Sacrifice Honored — 34

7. House on a Hill — 43

8. A Wise Economy	46
9. Moral Purpose	51
10. Sacred Architecture	58
11. Garden of Stone	64
12. J. Warren Keifer	69
13. Fade to Black	76
14. A Great Loss	82
15. Leveling the Field	94
16. A New Era	102

17. Stone Stories   107

18. Roads Less Traveled   116

19. Biographies   129

I cannot say, and I will not say,
That he is dead; he is just away.
With a cheery smile and a wave of the hand,
He has wandered into an unknown land.

—James Whitcomb Riley

PREFACE

Cemetery, Greenmount Cemetery and the venerable "Old Graveyard" known as Demint Cemetery (located on present-day Columbia Street) are commonly referred to as "cities of the silent" or "bivouacs of the dead." Unearthing their rich history and significance challenges visitors, first and foremost, to revisit the notion that cemeteries are merely resting places for those who have gone before. Consider instead the unique revelation that graveyards are places of *living* history, sustaining the lives of the deceased through memories rekindled, contributions recalled and stories retold.

I am convinced that no one dies in vain, believing that everyone, sooner or later, fulfills an intended, earthly purpose. Whether buried in Greenlawn, Demint or that glorious, pastoral retreat known as Ferncliff, rich or poor, well-known or not, Springfield's dead all once enjoyed a small sphere of influence among those who loved and befriended them. They were important to family members, important to friends, and remain important to us now. They are history's messengers, and their stories live on.

It's interesting, the manner in which we celebrate the lives of those who've passed into eternity. We travel to funeral homes, stand in receiving lines to offer condolences, then caravan to area cemeteries. As we park our cars and walk to nearby burial sites, we notice holes in the ground draped with velvet, engulfed by the beauty and fragrance of flowers. Tents shelter rows of chairs placed for the comfort of bereaved family members. Loved ones gather and listen as pastors deliver their final words before the deceased are returned to the ground. Once we leave these places of grief, we return to our homes and reflect, for a time, upon the impact each loss has upon families and friends. But the intensity of our attention is fleeting. Eventually, we return to our lives. This is the circle of life. This is our way. For as French aviator and writer Antoine de Saint-Exupery wrote, "On a day of burial there is no perspective—for space itself is annihilated. Your dead friend is still a fragmentary being. The day you bury him is a day of chores and crowds, of hands false or true to be shaken, of the immediate cares of mourning. The dead friend will not really die until tomorrow, when silence is round you again. Then he will show himself complete, as he was -- to tear himself away, as he was, from the substantial you. Only then will you cry out because of him who is leaving and whom you cannot detain."

Does anyone linger, as I often do, upon what death is like for those left behind? My own mother has been gone for several years now, but my feelings of loss persist. I find, in fact, that I miss her now more than ever before. With the passage of time, memories of her grow more and more vivid. Pictures scattered throughout my family's house preserve her vibrant smile, convey the fullness of her life. Her countenance is one of pure joy.

My father just recently passed away as well. Their ashes now fill a mausoleum urn inside Ferncliff Cemetery. As I approach the niche to visit, I sometimes talk to Mom and Dad, wondering whether or not they can hear me ... whether they recognize or know that I'm there. It's often said that the deceased "look upon us from heaven," watching us experience life in their absence. I wonder, though, if that's truly the case. While I have my doubts, it's a fanciful thought. After all, consider Hollywood's vision of death, where ghost-like characters, seen or unseen, visit the living. *A Christmas Carol. Ghost. It's a Wonderful Life.* Entertaining fare, to be sure, but alas, they *are* only movies.

To experience the lives of the dead, we need only to keep their memories alive in our hearts and minds ... to simply tell others about them. And therein lay the purpose of this book. In celebrating the lives of men and women who've gone before, contributing significantly to Springfield's growth, we discover the local ancestry that delivers context to our modern lives, forever influencing our agriculture, our industry and our society.

May they rest in peace, never forgotten. For as Cicero concludes in *Orations-P.*, "The life of the dead consists in being present in the minds of the living."

—*Paul W. "Ski" Schanher, III*

BEAUTIFUL *Ferncliff*

Sawyer Monument in Mount Auburn Cemetery

Rural Cemeteries

THE 'STORY' OF A MOVEMENT

Lives are commemorated, deaths are recorded, families are reunited and love is undisguised. This is a cemetery. Communities accord respect, families bestow reverence, historians seek information and our heritage is thereby enriched. Testimonies of devotion, pride and remembrance are carved in stone to pay warm tribute to accomplishments and to the life—not the death—of a loved one. The cemetery is homeland for family memorials that are a sustaining source of comfort to the living. A cemetery is a history of people—a perpetual record of yesterday and sanctuary of peace and quiet today. A cemetery exists because every life is worth loving and remembering—always.

—"This is a Cemetery"
Author Unknown

Ferncliff Cemetery

Cemetery of Mount Auburn, engraving by W.H. Bartlett 1839

Modern cemeteries, beautiful "gardens of our dead," are earning recognition and respect beyond their more practical roles as simplified places of burial. These ornate "cities of the silent" are products of comparatively recent times. Our forefathers were traditionally laid to rest inside church yards, small family cemeteries, community graveyards and municipally owned cemeteries. Today, although many of these historic burial places are sadly neglected, modern, perpetual-care cemeteries are ushering in a new era. More than mere places of burial, they offer thoughtfully designed, park-like settings where the living go to remember loved ones who've passed—places of quiet beauty and peace where flowers, trees and blooming shrubs harmonize with gentle slopes, winding drives and meticulously manicured lawns. Comforted and inspired by their quiet dignity, visitors return again and again. (Fig. 1)America's rural cemetery movement began in 1831 at the consecration of Mount Auburn, a 175-acre site near the Charles River on Massachusetts' Cambridge-Watertown border. The revolutionary idea of sculpting garden-like cemeteries from their natural environment inspired designers to envision weekend walks among the monuments. As the need for additional burial space developed, landscape architects incorporated public-park features into garden cemetery design. A new era was born.

An overview by the United States Department of the Interior, National Park Service (preserved on a National Register of Historic Places Registration Form) concludes, "Mount

Auburn provides striking visual evidence of the transformation in attitudes toward death and commemoration occurring in the nineteenth century. Throughout the colonial period, graveyards were barren, purely functional places for the disposal of the dead, displaying stern warnings about man's postmortem fate. Poorly maintained, Boston's burial grounds were overcrowded and full of gaping graves.

"Mount Auburn was a radical shift in taste. It offered a peaceful, serene location for the burial of Boston's dead in perpetuity, an innovative post-Revolution concept. Even the use of the term "cemetery" to refer to ground burial was unprecedented in the American vernacular. Derived from the Greek word for 'dormitory' or 'place of sleep,' it indicated new notions of death stemming from Enlightenment philosophy. Rather than depicting the horror of death, Mount Auburn's monuments and picturesque landscape were designed to provide solace and comfort to the bereaved, emotions now highly acceptable due to nineteenth-century liberalized religious attitudes.

Yarrow Path, Mount Auburn.

"In marked contrast to colonial burial grounds, the cemetery was also intended to provide a place for the commemoration of the lives and works of the deceased. Concerned with shaping a common celebratory history for the new nation, Mount Auburn's founders wanted monuments erected to reflect the accomplishments of notable individuals. This 'cult of heroes,' accompanied by appropriate inscriptions, was intended to provide instruction and enlightenment to future generations."

During Mount Auburn's dedication as America's first rural cemetery, United States Associate Supreme Court Justice Joseph Story (1779-1845) initiated the movement by referring to burial sites as "places of repose." Considered a brilliant legal mind and the youngest individual ever to hold a U.S. Supreme Court seat (1811-1845), Story was appointed at age thirty-two by President James Madison. He became Harvard's first Dane professor of law in 1829 and, during his thirty-four years on the bench, authored numerous influential works, among them a three-volume commentary on the U.S. Constitution (1833).

As Mount Auburn's first president, Story publicly questioned the early American practice of "depositing the remains of our friends in loathsome vaults or beneath gloomy crypts and cells of our churches, where human foot is never heard," describing the traditional graveyard of his day as "set apart from life, a narrow confine cut off from communication with its surroundings, walled in only to preserve them from violation."

Subsequent cemeteries would, instead, become "frequent resort(s) for the living, who would commune with nature as a way of finding life in death."

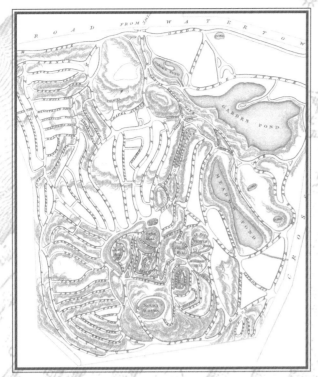

"Plan of the Cemetery of Mount Auburn" by Henry A.S. Dearborn.

Nominated for inclusion on the national register in 1975, its National Park Service overview finds that "Mount Auburn's most unique resource is the cemetery's own landscape, a combination of topography, water bodies, avenues, paths, living plants and historic monuments, buildings and other structures While it has seen great changes over 170 years, the original design of the rural cemetery is largely intact, and its historical integrity and origins are evident to visitors. Mount Auburn Cemetery preserves a remarkably illustrative chronicle of American landscape design, attitudes toward death and commemoration, aesthetic and spiritual values, material culture and changing technology.

"The cemetery's overall design was to serve as an example, consoling the bereaved and inspiring visitors to contemplate their mortality in an uplifting spirit of melancholy. Just as John Winthrop and the Puritans intended to make Boston a model community for the living, Mount Auburn's founders set out to create another model—a 'silent city on a hill' for the dead." Its outer gate quotes Ecclesiastes 12:7: "Then shall the dust return/ To the earth as it was/ And the spirit shall return/ Unto God who gave it."

Nineteenth-century rural cemetery planners believed cemeteries should be located outside city limits, etched into the serene countryside. Excitement brewed with the dedication of each new cemetery. Schools closed, businesses shut down, and household cares were laid aside as entire towns were encouraged to attend their dedications. Many residents devoted entire days to the solemnity of each occasion.

Establishment of Mount Auburn Cemetery provided the blueprint for rural cemetery dedications to follow: a procession of dignitaries, an opening prayer and a recessional. Noted Story, "The services will consist of music, prayer, an oration, a formal declaration setting apart the grounds to burial uses, and the Apostolic Benediction."

During the dedication of Gettysburg National Cemetery, golden-tongued orator Edward Everett announced, "Standing beneath this serene sky, overlooking these broad fields now reposing from the labors of the waning years—the mighty Alleghenies dimly towering before us, the graves of our brethren beneath our feet—it is with hesitation that I raise my poor voice to break the eloquent silence of God and nature."

Words such as Everett's convey the reverence

of such occasions. He and others painted masterful oral portraits that captured particularly poignant moments in time.

"All around us there breathes a solemn calm, as if we were in the bosom of a wilderness," Story said during his Mount Auburn address on September 24, 1831. "—Ascend but a few steps, and what a change of scenery to surprise and delight us. We seem ... to pass from the confines of death, to the bright and balmy regions of life."

The National Park Service credits Mount Auburn for its profound influence "upon nineteenth-century attitudes about death, burial and commemoration"—an aesthetic approach that was widely imitated. Within fifteen years, nine major cemeteries were similarly fashioned. The rural cemetery movement reached the Mississippi by 1849 and the west coast by 1863, and inspired not just cemetery architecture, but the nation's first public parks and picturesque suburbs.

Like Story, Scottish horticulturist and Gettysburg National Cemetery designer William Saunders believed a cemetery's prevailing expression should be that of "simple grandeur." His "Remarks on Design" in the *1865 Report of the Relations to the Soldier's National Cemetery* emphasized the following: "Simplicity is that element of beauty in a scene that leads gradually from one object to another, in easy harmony, avoiding abrupt contrasts and unexpected features. Grandeur, in the application, is closely allied to solemnity. Solemnity is an attribute of the sublime. The sublime in scenery may be defined as continuity of extent, the repetition of objects in themselves simple and commonplace.

"We do not apply this epithet to the scanty tricklings of the brook, but to the collected waters of the ocean. To produce an expression of grandeur, we must avoid intricacy and great variety of parts; more particularly, we must refrain from introducing any intermixture of meretricious display or ornament.

"The disposition of trees and shrubs is such as will ultimately produce a considerable degree of landscape effect. Ample spaces of lawn are provided; these will form vistas, as seen from the drive, showing the monument and other prominent points. Any abridgment of these lawns by planting, further than is shown in design, will tend to destroy the massive effect of the groupings, and in time would render the whole confused and intricate. As the trees spread and extend, the quiet beauty produced by these open spaces of lawn will yearly become

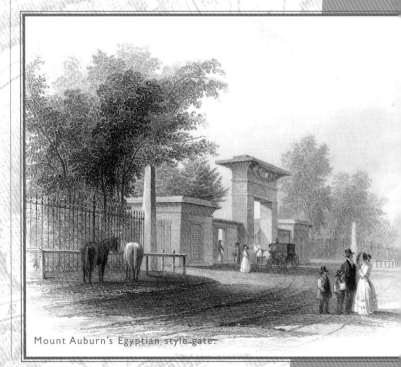

Mount Auburn's Egyptian style gate.

more striking. Designs of this character require time for their development, and their ultimate harmony should not be impaired or sacrificed to immediate and temporary interest."

Gettysburg's success is, interestingly, Ferncliff's success. No cemetery, after all, should have the beauty of its natural landscape blocked by monuments. Rather, conscientious designers champion a "pleasing landscape and pleasing ground effects."

As *Lincoln at Gettysburg* author Gary Wills writes, places of the dead "must be made school(s) for the living." A casual drive or stroll through Springfield's Ferncliff Cemetery, ground patterned in the rural tradition, emphasizes the living within the natural beauty of its design. All 260 acres are adorned with a rich variety of trees and ornamental shrubbery. Water cascades gently down the moss-covered cliffs that line its Plum Street entrance. Picturesque Kelly Lake and its stately swans invite visitors to linger just a while longer. The very definition of a "place of repose," Ferncliff is, indeed, alive—a locality for the living.

McGilvray Mausoleum, Section R

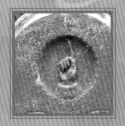

Pioneer Spirit

TILLING GOD'S ACRE

I like that ancient Saxon phrase, which calls, The burial-ground God's Acre! It is just; It consecrates each grave within its walls, And breathes a benison o'er the sleeping dust.

—"God's-Acre"
Henry Wadsworth
Longfellow
(1807-1882)

George Rogers Clark

Ferncliff Cemetery's official dedication on July 4, 1864 unleashed a civic pride of unusually large proportion in the modest yet enterprising town of Springfield, Ohio. Locals hailed the property's rural, state-of-the-art magnificence, grateful for the long-overdue respect their city was finally paying its dead. "The fact is, our cemetery park, as a whole and in detail, is one of the finest things of its kind in America, and there is certainly nothing like it in Europe," opined *The Springfield Republic* on Oct. 11, 1886. "There are finer parks, as a matter of course, but no cemetery is anything like it. We hope our citizens will make themselves familiar with the full value of this splendid local feature, and make liberal provision for still further increasing its attractiveness. Nowhere are the refinement and taste and true public spirit of our people more credibly illustrated than in Ferncliff."

To fully appreciate the cemetery's lasting architectural and historical significance, a trip into Clark County's past is essential.

During the August 8, 1780 Battle of Piqua—the American Revolution's largest frontier fight west of the Allegheny Mountains—Native American wilderness that saw only the occasional French trapper or British trader was opened to an increasing number of white settlers. With an army of 1,000 Kentucky soldier recruits marching on the command of General George Washington, Colonel George Rogers Clark drove across the Ohio River near where modern-day Cincinnati is located, purging Ohio's southwest territory of Indians to further the development of pioneer settlements. Commissioned by Clark to serve as a guide was Simon Kenton, a well-known Indian fighter and frontiersman. Advancing through the vacated Indian village of Chillicothe, today known as Oldtown and located just outside of Xenia, Clark and his troops crossed the Mad River and defeated 300 Native Americans in Piqua, a large village of Shawnee, Wyandot, Mingo and Delaware tribes. Suffering minimal loss of life, perhaps no more than twelve men, Clark's unit made the area somewhat safer for white settlement ... but the downturn in violence didn't last long. Indians killed John Paul and his family in 1790, not long after they settled near modern-day New Carlisle, prompting Washington to commission Gen. "Mad Anthony" Wayne to secure the area as a pioneer stronghold. Utilizing 2,000 men from the Kentucky regions, he led an attack in 1794, soundly defeating the Indians at Fallen Timbers. The 1795 Treaty of Greenville followed, returning a sense of calm to white settlements.

Four years later, in 1799, Simon Kenton, John Humphreys and frontiersman James Demint of Loudon County, Virginia, each established pioneer homes in the area. Having previously fought with Clark at the Battle of Piqua, Kenton returned with Humphreys and a handful of other families to settle, establishing a community located west of present-day Springfield in the front yard of what is, today, the Ohio Masonic Home, overlooking the confluence of Mad River and Buck Creek (where an Ohio Edison plant now stands). A springhouse marks what is believed to be the exact location of the Kenton-Humphreys settlement—fourteen cabins and a blockhouse to serve as protection against Native Americans.

Demint, meanwhile, traveled with his family from Kentucky. Today known as Springfield's founding father, Demint contracted with John Cleves Symmes, a Cincinnati real estate developer, for 640 plots of land located along Lagonda (Buck) Creek and built a cabin along the creek's northern edge, near where the William Toy Board of Education building now stands. The modest cabin is believed to be the first such dwelling built within Springfield's modern-day city limits.

IF YOU BUILD IT, THEY WILL COME

In 1801, Demint envisioned the establishment of a town, and with guidance from surveyor John Dougherty, a native Virginian who accompanied Demint from Kentucky, laid out a plat for the village of Springfield. The plat was later dated and signed by Demint on September 5, 1803, and recorded in Greene County on September 13, 1804. The budding village consisted of ninety-six lots averaging 99x198 feet and extended from the middle of High Street on the south to Buck Creek on the north, to Spring Street (known then as East Street) on the east, and westward to Fountain Avenue (once known as both West Street and Market Street). The lots were grouped around a public square near where the Springfield Post Office stands today. The public square at this time was 260 feet wide and 460 feet long.

Simon Kenton

Griffith Foos arrived in Springfield from Franklinton, Ohio, west of present-day Columbus, in 1801, accompanied by his family and several friends. They followed Indian trails westward, stopping for rest atop ground fed by an underground spring. They happened upon the Mad River and, later, Lagonda Creek, following its banks to a cabin owned by James Demint. He offered his unexpected visitors several days' rest and the prospect of low-priced land, sharing with them his plans to establish a town. Interested, Foos retrieved his family from Franklinton and quickly returned. He not only assisted in laying out Springfield's first plat, but bought the first lot that Demint sold—on the southwest corner of

Looking south towards the Civil War mound.

Main and Spring Streets. There, he built his home (the town's first), earning the nickname "Father Springfield" among local historians. The double-log cabin was later transformed into Springfield's first inn.

As James Demint laid out the plat, three lots of 1.49 acres located west of Center Street and Columbia Streets were reserved as burial ground. Until 1844 this property was shared by the community, becoming the town's first cemetery connected to civilian life. Previously, mound builders and other Native American tribes buried their dead in mounds scattered throughout Clark County. Enon's Adena Mound remains perhaps the area's finest example of that ancient tradition.

Upon mutual agreement, Demint and fellow pioneer Robert Rennick divided the land in and around Springfield, with modern-day Plum Street providing a line of demarcation. Demint chose the eastern half containing the newly platted town, while Rennick selected the undeveloped west. Along with his wife, eight children and father, Robert Christie arrived in 1817 from Washington County, Vermont. Eventually tiring of Springfield's city life, he and his family retired to a farm on Rennick's land, where they prospered for quite some time. The Christie home became the center of many social events, with the family providing hospitality to neighboring farm families and local citizens until Robert Christie's death at age forty-seven in August of 1822. His 87-year-old father, Deacon Jesse Christie, followed five months later in January.

Shortly thereafter, Henry Bechtle purchased the 150-acre Christie farm and moved there with his wife, five sons and three daughters, taking official possession of the home in 1826. On the farm sat a prosperous grist mill established around 1806 or 1807 and operated until 1835.

Bechtle died on February 9, 1839, but the importance of his property becomes evident as Ferncliff's story unfolds.

During this time, Demint Cemetery served as a public burying place for deceased locals. In 1803, the fourth person laid to rest there was James Demint's first wife, Elizabeth Greeley Demint. Upon his death in Urbana in 1817, Demint was buried there, too, along with his second wife, Nancy.

Due to insufficient records the site's initial three burials remain unknown. However, surviving paperwork does indicate that seven Revolutionary War soldiers were buried in Columbia Street's hallowed ground, including Boston Tea Party rebel Elijah Beardsley and James Kelly, grandfather of noted manufacturer Oliver S. Kelly. Two of the seven have since been removed to other cemeteries.

Columbia Street Cemetery was not surveyed, laid out, or dedicated until 1814, when Demint expanded Springfield with a layout known as Demint's Addition West of Mill Run (preserved in Vol. 4, Page 3 of Clark County's plat records, Lot Number 103). Despite its historic role as the public burial ground of Springfield's founding fathers, Columbia Street Cemetery ceased to be used around 1844. Once sadly neglected, the notable property became a disgraceful tangle of briars and brush, filled with discarded bottles and assorted trash before the local Daughters of the American Revolution (DAR) prompted city officials to clean, beautify and maintain the ground—a fitting tribute to the pioneering men and women buried in "God's Acre."

Its oldest stones have long since disappeared, and most of those that remain have been rendered illegible from decades of weathering. While no proper records of ownership or burials survive, Columbia Street Cemetery and her delicate history remain a lasting testament to Springfield's forefathers.

BOOM TOWN

Springfield's growth can be traced to the early 1840s, when advances in farming and manufacturing expanded local population and, in turn, the need for a larger, more suitable cemetery. James Leffel, the city's first inventor, developed the use of water power by constructing a mill complex along Buck Creek—one that amply served the needs of a growing village. William Whiteley, meanwhile, perfected Cyrus McCormack's reaper, leading Springfield into the "Golden Age" of the 1860s as Ohio's third-largest city.

The subject of a suitable burial ground for Springfield's dead had, however, been brought before city council much earlier ... in 1842. Land was selected where Wittenberg University now stands, but following several burials, was awkwardly transferred back to the school. The college-grounds cemetery of Woodshade served, for a while, as a city burial place, with one of the first interments being that of Ezra Keller, founder and inaugural president of what was then known as Wittenberg College. City council members eventually made arrangements to purchase a tract of land located east of the city on the National Road, between Main and High streets, in an area known as Greenmount. The beautifully wooded, rolling elevation consisted of 12.5 acres—a site later transferred to the city in 1845 by Cyrus Armstrong and deeded on September 2, 1845 for $1,256. A plain but substantial home was constructed at a cost of $1,000, and the first burial took place in December of 1844.

But with Springfield fast becoming a boom town, rapid city growth soon made evident that after twenty years of use, Greenmount Cemetery was no longer sufficient for proper burial of the dead. With this concern in mind, city council members met and noted in their minutes: "The demand for burial lots (has) so far exceeded the probability of the city council to supply, which partly holds control of the burial grounds, that the members (feel) compelled to take action on this subject, and either procure additional land for the purpose, or else attract the attention of the citizens to the subject, which should lead to action.

"Accordingly, at our regular meeting, the city council, held Tuesday evening, June 23, 1863, Mr. William Warder offered a resolution requesting all citizens interested in the subject to meet with the city council on Saturday evening, June 27, at the courthouse, to take sound action in reference to the purchase of a new cemetery."

Additionally, the following appeared June 26, 1863 in The (Springfield) Republic newspaper: "The city council invites the citizens to meet its members at the courthouse tomorrow evening for

A booming Springfield in 1854.

the purpose of consultation upon, and discussion of, the plans, which are proposed for the purchase of a new cemetery. The council has taken a timely step in the right direction, and we trust that our leading citizens will respond to this invitation. The new cemetery must be provided. It only remains to decide how and where.... The object of the meeting demands a general attentiveness of citizens who possess public spiritedness and have regard for the permanent prosperity of our city."

Obtaining input from concerned citizens in attendance, council decided that the "purchase of a cemetery could not be much longer delayed." But key questions remained. How should the new purchase be made? By council? By the citizens? By an association of citizens through which council is represented? The last-mentioned plan seemed best. Grounds were offered at varying prices. On a farm south of Selma Pike, land could be obtained for $226 an acre. Land opposite Greenmount Cemetery, meanwhile, was offered at $250 an acre. A tract across from Buck Creek, located on the Bechtle Farm, could be had for $175 per acre.

Decisions had to be made. Resident George F. Frey offered a resolution: that the "organization of a cemetery association is, in the highest degree, central to meet the needs of the community."

Frey called for prompt action, urging, there "is no community need more imperative than a suitable place for its dead. Hundreds of families have no lots in Greenmount. That (is) sufficient reason for a new cemetery. It (is) important to have a beautiful cemetery retired from the noise and dust of the city, and it (is) important that it should be managed." He did not believe that council, as a body, was competent, "for it had not managed Greenmount as public taste and public needs required."

William McIntire, a first ward councilman, agreed, adding, "The city (has) never planted a shade tree in Greenmount. (Has) never grated a walk." He encouraged citizens to "take hold of the enterprise with public spirit and procure suitable cemetery grounds. Committee chairman Chandler Robbins believed, too, that the enterprise "should be in the hands of an association," and that it was "a disgrace to Springfield that it did not have a suitable cemetery." He had, in fact, "often been mortified when strangers asked him, 'Where's your cemetery?'"

Frey's novel idea was to appoint three citizens from each ward and four members of city council to take the subject of a new cemetery into consideration. Samuel A. Bowman, a prominent attorney, remained of the decided opinion that only a committee of citizens should be appointed. He offered the following resolution: "That a committee of three persons, one from each ward of the city, be appointed to report a plan of organization for a cemetery association, and also facts and statistics concerning the operation of associations in other cities." Led by Gen. Samson Mason, a war veteran who practiced law, others, too, came to believe that the community's goal may be best met through means of an association.

As a matter of courtesy, Frey thought the city council should be officially recognized on this preliminary movement, and his opinion coaxed Bowman to withdraw his resolution. The original motion was unanimously adopted: the city council president shall appoint the committee members from council, while the meeting chairman shall appoint its citizen members. Bowman reported on July 20, 1863 that "the most practical manner in which (a cemetery association) can be accomplished is to receive a subscription, in the nature of a loan, from each member of the association."

He suggested the following obligations: "We, the undersigned, desirous of organizing a corporation under the act of the general assembly of the State of Ohio, providing for the organization of cemetery associations passed February 24, 1848, do hereby subscribe the several sums next to our names for the purpose of buying and embellishing grounds to be used forever as a rural cemetery near this city. These subscriptions (are) to be binding whenever a cemetery association is organized under said act of 1848."

"The subscribers are to pay the association in the nature of a loan, to be repaid with interest out of the proceeds of sales of burial lots, under which rules and regulations as the corporation may prescribe. Our committee further reports that the act of 1848 limits the capacity of the corporation to contract debts to the sum of $10,000, which sum we recommend as a limit of subscriptions to the foregoing obligations, and that no subscriptions to the foregoing obligation less than $100 be received."

The Springfield Cemetery Association was born.

Scenic Kelly's Lake in autumn.

A Civic Duty Fulfilled

THE BIRTH OF FERNCLIFF CEMETERY

*Cherish it, my friend;
preserve it, my friend,
For stones sometimes
crumble to dust
And generations of folks
yet to come
Will be grateful for your trust.*

—Thelma Greene Reagan,
National Tombstone
Project

The newly formed Springfield Cemetery Association would offer a timely solution to the pressing problem of where to properly bury the city's dead. A sense of urgency propelled the committee in to swift action. Committee Chairman Chandler Robbins authorized several people to visit cemeteries in other cities, and to submit reports on their size and expenditures. The pressing nature of the problem was reinforced when city clerk James Cummings noted that only eight vacant lots remained in Greenmount Cemetery, and warned that within 60 days, the city would likely have no further room for burials. A public meeting was scheduled to discuss how to secure subscriptions and inaugurate a new cemetery movement.

On August 3, 1863, Robbins reported that Springfield's citizens were concerned with the lack of burial space at Greenmount, pleased with the cemetery movement, and, as was the committee, "generally surprised—greatly surprised—to learn how imminent and pressing the necessity for a new burial place (is)." He stated, "It is well known to most of our citizens that there are well-founded objections to the present burial place, having reference to the nature of the soil and the absence of taste in laying out the grounds. So unanimous, however, in our own community is the opinion of the want of a new cemetery of rural character, laid out in pleasant drives and improved with a view of developing landscape beauties. ... So universal is a wish in the savage, no less than in the enlightened, to surround the graves of the dead with scenes and objects indicative of honor, respect and affections that it well may be regarded as an instinct of our nature."

Discussion ensued regarding who should control the cemetery. Referring to city council's poor management (and lack of attention to) Greenmount, Robbins concluded, "it is evident ... that they have neither the time nor the disposition to take charge of the new enterprise, requiring so large an amount of attention and their time. It is well known that the members of the council desire to be relieved from a charge which might give them annoyance, and would certainly impose on them a considerable increase of unpaid labor. Your committee has found that in most of the towns in the state where there are rural cemeteries, they are under the management of associations incorporated under special acts or under the general law. It would seem to be the general experience that an association of persons organized for the special purpose, and for no other, of conducting a business pertaining to cemeteries will be much more likely to conduct it successfully, and to the satisfaction of the public."

The committee chairman recommended that subscriptions be opened to the public in the amount of $300 each, with total capital stock not to exceed $10,000.

WE, THE PEOPLE

The Honorable Samson Mason, a famed War of 1812 attorney, proposed that an organizational plan for a cemetery association be approved, and that cemetery loan subscribers organize themselves into a corporation. He requested that Springfield city council aid the proposed new cemetery association by purchasing a portion of the grounds.

On August 25, 1863, a cemetery association assembled from citizen interest was formed. Reporting to initial cemetery fund subscribers, Bowman proclaimed, "We form ourselves into a cemetery association."

Election of officers proceeded quickly. Dr. Robert Rodgers, S.A. Bowman and D. Shaffer were elected to three-year terms as trustees, G.S. Foos and Chandler Robbins were elected to two-year terms, and William Warder and John Ludlow were tapped for one year of service. David Cooper was selected clerk.

Ludlow submitted a report detailing available tracts of land and their purchase price. Concerned citizens stepped forward to offer portions of their land for sale. Property owner William Rodgers offered forty-one acres off the east side of Urbana Road, near the first turnpike gate, at $112 per acre. James Halsey, an area farmer with land near the adjoining community of Lagonda, pitched eighty-two acres at $7,000—roughly $85 an acre. Others proposing land deals included Henry Bechtle, who was willing to part with seventy acres west of Wittenberg College for $150 an acre. Land was also offered along the National Road, across from Greenmount Cemetery—fifty-five acres at $200 each.

Known today as "Father Ferncliff" for his lifelong devotion to Springfield's rural cemetery project, Ludlow was elected to the first of his many terms as board of trustees president. Foos was elected treasurer. Together, they were authorized, subject to association approval, to purchase seventy of 230 acres owned by Bechtle's widow and heirs for $100 an acre.

Each of the association's cemetery fund subscribers was required, within twenty days, to pay to the treasurer "an installment of 33-1/3 percent of the subscription." Treasurer Foos issued certificates to each subscriber, their investments payable at six percent interest from the sale of burial lots. A contract with the Bechtle heirs was eventually resolved on September 3, 1863, stating, in part, that the sixty to seventy acres of land "in the northeast quarter of Section 5, Township 4 and Range 7, Springfield Township, Clark County, Ohio, be ratified and approved, and that the board of trustees be authorized to purchase as much of the said land, not exceeding 75 acres, as they may deem expedient."

Executive committee members Warder and Foos oversaw the contracted removal of dead timber and assorted living trees to prepare the grounds for layout and design of a rural cemetery. Joining a handful of other members, they traveled throughout Ohio, visiting and evaluating model cemeteries of the time. Civil engineer William Brown was hired to complete a topographical survey and map of the grounds.

Following Brown's survey, the cemetery association passed a resolution to officially purchase 70.8 acres of land from the widow of Bechtle. In October of 1863, Bowman and Warder were dispatched to Cincinnati, where arrangements were made to employ one or more landscape gardeners to lay out and beautify the grounds. Within a month, Cincinnati native John Dick was hired as the cemetery's inaugural superintendent at a salary of $500

The Buxton Monument, Section E.

and a winter's-worth of firewood annually. Arrangements were made for him to live as near the grounds as possible and be assisted in buying a house.

The Springfield-Cincinnati connection was perhaps best explained by the establishment of Spring Grove Cemetery in 1845. One of an early handful of cemeteries nationwide modeled after historic Mount Auburn, Spring Grove was, at the time, one of the country's leading rural cemetery descendants. Eventually, though, such designs proved too time-consuming and costly to maintain, giving way to the abbreviated "lawn park" style initiated at Spring Grove in 1855 by Prussian landscape artist and acting superintendent Adolph Strauch, whose influence quickly spread throughout the state.

According to the text of the National Parks Service Historical Site Nomination for Woodlawn Cemetery in Lucas County, Ohio, Strauch's increasingly popular "lawn park" designs served as "modification of the original rural cemetery movement as a response to problems that became apparent in early rural cemeteries. The landscape lawn plan, with monument only, was an effort to restrict patrons who decorated their individual lots with disregard to the general effect of the landscape. This threatened to detract from the landscape effect, which was a principle feature of the rural cemetery movement. Woodlawn's park-like appearance and careful attention to ensuring grave markers, tree plantings and roadways best compliment the landscape is an example of the movement to form a consistent whole. Strauch's idea was that the best effect is obtained 'when broad undulations of green turf prevail, adorned here and there with a noble family monument, and at proper intervals, shaded with suitable trees. Such lots, blending the elegance of a park with the pensive beauty of a burial place, confer on the whole a grace and dignity which can never be obtained in situations where every foot of ground is occupied with ornamental puerilities (sic).'"

Thus, the popularity of rural cemeteries, ironically, contributed to their eventual decline. The maintenance of a landscape design that included wild spaces was tremendously time consuming and expensive. A large workforce was necessary as combating overgrowth was a constant battle. In addition, rural cemetery lot owners had the right to design and plant their lots as they saw fit. The result was a cemetery that lacked any visual uniformity and was impossible to maintain.

The problems inherent in the rural cemetery movement led to the development of lawn park cemeteries, which married the attributes of landscape design, but imposed upon it a system of rules and regulations. The design stressed clearing the dramatic natural landscapes of yesteryear and manipulating the grounds into a natural-looking greensward. Trees, shrubs and other plantings were kept to a minimum to allow the play of sunlight over green lawn. The effect was one of restraint, both in the landscape and in monuments. Under Strauch's plan, lot owners lost the ability to fence their lots, send in their own gardeners, or add any plantings to their property. The cemetery was to provide service and care to the grounds as a whole, thus maintaining a unified landscape.

"Strauch applied his system to Spring Grove Cemetery in Cincinnati, Ohio starting in 1855. Spring Grove was 15 years old by this point and was already collapsing under the weight of its rural design."

With plans and vision well researched, the association christened Springfield's newest cemetery "Ferncliff" on April 4, 1864, appropriate given the wealth of ferns that arose from cracks in the moist, rugged cliffs along its Plum Street entrance. In June, Superintendent Dick was granted permission to lay out the grounds into lots, and preparations were made for their dedication and sale. A circular mound near the center of the property was set aside as a burial place for Clark County's deceased soldiers, who, honorably, were laid to rest without charge. A commemorative soldiers' monument was proposed to mark the site.

Official dedication of Ferncliff Cemetery followed on July 4, 1864, with the Honorable Samson Mason presiding. Exercises opened with celebratory singing, Reverend Dr. Joseph Clokey offered a prayer, and Reverend S. Scolvel delivered the dedicatory address. Reverend and trustee Chandler Robbins presented a historical statement regarding the association's origin and rules, with the Honorable Samuel Shellabarger, a prominent attorney and congressman, announcing plans for a soldiers' monument. Reverend Samuel Sprecher, Wittenberg College president, issued a dedication, and upon singing the doxology, Reverend Edward Root issued an event- closing benediction.

Lot sales began at three o'clock that afternoon, and an audience again assembled to participate in the bidding. John F. Chorpening was the highest bidder, paying $570 for the privilege of owning a prime site located in Lot 1, Section B. John Foos selected Lot 1, Section A with the second choice, paying $252. Also of note, bidder William Warder selected Lot 13, Section C with the 12th choice for $177 and purchased several additional lots as well.

The first interment was made on the grounds on June 21, 1864. According to a report of the clerk on November 23, 1864, the event marked the burial of a young boy named George Clemens, who died of spotted fever at the tender age of eight years, eleven months and twenty-four days. The total number of dead interred at this time was sixty-six (which included one removal from Cincinnati's Spring Grove, five from the college grounds, and twenty-seven from Greenmount). Ninety-nine lots were sold totaling $10,240.20. Ludlow enthusiastically reported that Spring Grove Superintendent Strauch believes young Ferncliff Cemetery "possesses more natural advantages for improvement than any other piece of ground of the same extent he has ever seen, either in Europe or this country." Ludlow added, "This would deem no ordinary recommendation, coming as it does from a gentleman of experience and taste in the art of landscape adornment. It's one of the most beautiful places for a rural cemetery."

During that same year, an important addition to the cemetery was completed at a cost of $914.26: construction of a cave to serve as a holding place for those who die during the winter months (when the ground is too frozen for digging). The cave was named "Machpelah," from Genesis 23—the story of Abraham being called to God—to leave the land of Ur with his wife, Sarah, and travel to Canaan.

According to Scripture, Sarah dies after 127 years, and Abraham, stating that he is a stranger, a sojourner in the land, asks the sons of Heth for a burial site. They offer him the choicest of graves, and Abraham chooses the cave of Machpelah, offering full price for the privilege of burying his beloved Sarah there. Ephron, who owns the cave, initially rejects payment for the burial site, saying, "I give you the land and the cave that is in it, with these people as witnesses. Bury your dead wife (New American Standard Version Bible)." Eventually, Ephron accepts Abraham's full payment of 400 shekels of silver (200 pounds of silver by today's standards). The field and cave are thus deeded to Abraham.

Machpelah cave.

Records do not indicate who etched "Machpelah" above the door of the Ferncliff Cemetery cave, but the structure remains aptly named.

STEADY GROWTH

Other improvements during this period included a footbridge across Buck Creek, and "duly noted in the design of the plan of improvements," was the association's indebtedness "for the gratuitous services of Mr. Adolph Strauch of Spring Grove Cemetery in Cincinnati, who has made two visits to Ferncliff for that purpose."

Ludlow's report of August 15, 1865 surmised "the whole grounds should be enclosed, so as to protect it from the encroachments of man and beast. No entrance should be had except the gateway, where a suitable residence should be built for the superintendent, (and) where he (and) his family could see all (who) go in and all (who) come out. Upon this building should be affixed a bell to announce the approach of funerals, and to call the superintendent to his office for business.

"A stable with accommodations for several horses and carts shall be erected, and near the residence of the superintendent should be a piece of ground suited to the cultivation of choice trees and shrubs for sale to lot owners. When the grounds are once well sodded with grass, the superintendent would furnish hay for his horses, and the mowing and care of the grounds. Suitable roads and crossings to Lagonda should be made to accommodate all parts of the city, and with easy and direct passages to the entrance gate. These things might all be accomplished in another year. At the dedication of the grounds on the Fourth of July last, the board of trustees concluded to set apart the grounds within the circle around the Indian mound as a place for the burial of soldiers from Clark County, who die in the services of country during the War of the Rebellion, and as a fit place to erect a suitable monument in their memorial."

Ludlow further reported, "much difficulty and considerable loss of labor has occurred from frequent washing of the avenues by rain, but a system of drainage is now being adopted, which is thought will prevent further difficulty of the kind. ... Still put to great inconvenience in not being provided with a near and easy entrance to the grounds. Funeral processions are obliged to trespass upon private property and cross the creek upon a temporary footbridge kept up by the association, or ending the journey of several miles by the northern road. I'm happy in being able to state that the city council (has) now before (it) the subject of a bridge and road, and several of its members have taken hold of the subject with sufficient zeal to secure success.

"The cemetery grounds are yet without the necessary protection against the encroachments of careless and designing intruders, and until suitable picket fence encloses the grounds, with the residence of the superintendent at the entrance gate as a sentinel, it would seem almost useless to try to enforce laws or establish rules for the protection of the place. Many of our citizens, both in the town and the country, have shown their interest in the enterprise of purchasing lots at the liberal prices, thus furnishing means for improving the whole. Yet, a much greater number of persons must come forward and give their assistance before the company will be able to complete the necessary improvements, such as the erection of a residence for the superintendent at the gate, of an enclosure that may protect the premises against the encroachments of man and beast, and until this is accomplished, the officers of this society should not slacken their labors."

A Civic Duty Fulfilled

Autumn in Ferncliff Cemetery.

Home is where the heart is

HOUSING THE SUPERINTENDENT

A house, we like to believe, can be a noble consort to man and the trees. The house should have repose and such texture as will quiet the whole and make it graciously one with external nature.

—"The Natural House"
Frank Lloyd Wright
(1867-1959)

Predictably, indebtedness became problematic during the early stages of Ferncliff development. In providing a quality, rural cemetery for the Springfield community, assessments of $100 or more were often necessary to pay off accumulating debt. As reported in the annual meeting of August 28, 1866, John Ludlow brought to association members' attention that the project's last annual installment of $1,750, with interest, "becomes due on September 25." Additionally, clerk records showed a second debt of $1,000 for borrowed money. Undeterred by these financial challenges, Ludlow remained determined to steer his rural cemetery vision toward completion. It appears impossible, he said, to provide for the indebtedness, given the slow-paced sale of burial lots and the cost associated with maintaining the grounds and honoring the salaries of the superintendent and clerk.

In view of these circumstances, Ludlow advised the new board of trustees to "make an assessment upon the members sufficient to place the cemetery out of debt." Such action "would enable the association to employ all the income from the sale of lots to the necessary expenses until such a time as funds can be conveniently spared to refund the stockholders."

On September 21, 1866, a motion was resolved authorizing the president to borrow $1,200 from Lydia Lawrence, a local citizen interested in Ferncliff's growth, to pay the balance due Mrs. Bechtle for the purchase of the cemetery's property. Her debt is to be paid off at the end of one year, interest included. Though cumbersome at times, the association's ongoing practice of borrowing from the community provided an opportunity for citizens to actively participate in the development of Springfield's beautiful rural cemetery enterprise.

Because a quorum was not present at the 1867 annual meeting, thereby preventing the transaction of business, two years passed before another report was provided. In August of 1868, Ludlow reported that cemetery work had not progressed in as timely a fashion as some would desire. Yet despite this, proceeds from lot sales enabled the superintendent to make key improvements. Several avenues within the grounds were completed. Surveying, grading and sodding of newly developed sections continued the project's slow but determined pace. Such upgrading became necessary due to frequent rainfall washouts, requiring roads to be raised and bouldered to remain open and prevent erosion. Several fences were also replaced due to the area's frequent flooding rains, and all work was completed quickly to prevent further damage. Impressively, the association's debt during this productive period was slowly reduced.

Ludlow again passed his concern to the board of trustees regarding a superintendent's residence, remarking, "The interests of the association and that of the public suffer for want of a dwelling within the grounds for the superintendent. At present, his residence is unavoidably at the opposite side of the creek, where he is obliged to spend a portion of his time as necessary to the protection of the cemetery."

Curiously, although Ludlow asked that his name not be placed on the ballot for re-election

The Plum Street Gatehouse served as the superintendent's first residence.

as association president—he believed occasional leadership changes were healthy and good—he was nonetheless re-elected—a vote of confidence in a tired but devoted man that, ironically, remained unshakable until his death.

VISION IN MOTION

The year 1869 brought revolutionary change to Ferncliff. Trustees authorized president Ludlow and Superintendent Dick to purchase a portion of property owned by the Gwynn family. The property rested outside the cemetery's south entrance gate along the rock cliffs north of Lagonda Creek, and the purchase provided sufficient breadth for both an entrance gate and a feverishly sought superintendent's residence. The board of trustees reported on August 31, 1869 that receipts for the sale of burial plots were larger than in any previous year, and as a result, work on the grounds advanced in relative, healthy proportion. Ferncliff Cemetery was well on its way to becoming reality: a work of art brought to the assistance of nature.

The association secured 6.48 acres of Gwynn land for $225 per acre and six-percent interest, expanding Ferncliff to seventy-seven acres. There remained a noted preference for an entrance gate and residence constructed of stone, but limited resources made wood a wiser choice, the cost of which was at least $2,000 less than the best estimates at the time.

A recently completed bridge over Buck Creek (on a line with Plum Street) was expected to increase traffic to the cemetery gate. The work of opening Plum Street was cheaper than that of Yellow Springs Street, would more easily accommodate nearby Wittenberg College, and offered connection to St. Paris Pike. The board opposed extending Yellow Springs Street because it entered the grounds behind the cemetery gate and could cause abandonment of the only suitable location for both a gate and residence (built below the rocks in a direct line with the main entrance at the recently constructed vault). Yellow Springs Street also remained largely unsafe for funeral processions due to its long, high bank that ran along the south side of the creek to the top of the rocks to the north, where it emptied into the cemetery grounds. Such would require building a pair of gates and the employment of two gatekeepers. Perhaps more importantly, the beautiful drive through the college grounds, along the scenic creek and rocks, could not be experienced en route to this magnificent rural cemetery.

The board therefore passed a resolution on November 25, effectively communicating to city council its resistance to appropriating any part of the cemetery's grounds for extension of Yellow Springs Street. Board member George Frey followed with a motion on September 1, 1871 that "trustees devise some plan by which cemetery grounds can be better protected by issuing tickets to persons visiting the grounds on the Sabbath day." This motion carried. And against his modest and soft-spoken wishes, the humble Ludlow was re-elected president.

A mild challenge emerged when Lydia Lawrence requested payment for her loan of $1,200 to the cemetery's early development. Ludlow decided to pay off the loan by securing a second loan and borrowed the sum from First National Bank, indicating the association's financial hardships weren't entirely resolved. Ludlow voiced his concern regarding the risky practice of borrowing money to pay off existing debts, discouraged that lot sales couldn't yet keep the project out of the red. The group reluctantly borrowed $1,000 from John Snyder, a prominent citizen who, along with his brother David, eventually donated to the city, in 1895, 225 acres of farmland for creation of a park. Because another $1,000 might be needed to pay off the land's third and final payment when it came due, Ludlow believed "with the state of things before us, it becomes a duty of the members of the association to take early measures for an increase of this revenue and the extinguishment of this debt. The welfare of the society would doubtlessly be promoted by a change in this office to one who is able to give more of his time to the business of the association. I therefore earnestly desire that the society will choose a more efficient person to serve them."

Ludlow's plea, possibly born of fatigue or financial frustration, fell on deaf ears. The visionary Springfielder was held in such high esteem that board members re-elected him yet again as a strong, respected leader they viewed as having no equal. Humbly, Ludlow accepted their unanimous decision and continued as association president, and at the annual meeting of August 26, 1879, said it was no longer necessary for the society president to continue the tradition of presenting a written report on cemetery conditions and progress. He did, however, suggest that a committee be appointed by the board of trustees to "solicit funds for the erection of a water works in the cemetery, and also for such other improvements and adornments as may be grown out of the water supply, or such funds (as) sufficient for its work."

Importantly, he also questioned whether the location of the superintendent's house was a healthy one. Although the location was one of "rustic beauty and of great conscience to the superintendent—frequent spells of sickness in his family, and the loss of life, (have led) to the thought that a residence on the higher part of the grounds and further away from the dampening influence of water might—be more conducive to their health." Ludlow was referring to the events of October 1878 when one of the most tragic chapters in Ferncliff's history took place. It was at this time that Superintendent John Dick and his wife lost both their youngest and eldest child within ten days of each other to diphtheria. With this tragedy still weighing heavily on his mind, Ludlow willed trustees to take action and arrange for proper inquiry and examination of the residence by competent professionals. (A sad postscript to the story is that the following October, as the anniversary of the deaths loomed, Mrs. Dick, overcome with grief, drowned herself in Buck Creek and was found by one of her other children. John Dick's parents arrived shortly afterward from Scotland to help the grieving family, but within five weeks his father also passed away.)

Returning to the events of the annual meeting of August 26, 1879, it was at this time Ludlow experienced a sense of relief in the association's improved financial standing, announcing it was now "a source of great gratification to us all that the cemetery is free from debt. It should be our aim in the future to avoid all burdensome debt and make no expenses which cannot be paid out of the cash in the cemetery treasury."

No more borrowing to pay off debt.

S.A. Bowman offered the following resolution on August 25, 1880: "that the board of trustees be directed to set apart 20 percent of all the proceeds of sales of the lots hereafter, made and invested same as a permanent fund, and that the principle of the said fund not be used except by the orders of the corporation." His motion passed—and its significance becomes evident as Ferncliff's story unfolds.

Due in generous part to the donation of a one-horse mower by William Whiteley and a complete lawnmower by Phineas P. Mast (both local businessmen of rich renown), Ludlow happily announced that "the saving of labor has been increased and the expenses thereby lessened. 'Tis some satisfaction to us all that the cemetery is out of debt with $394.50 of cash in the treasury."

On August 16, 1881, the association's annual meeting date officially moved from the third Tuesday in August to the first Tuesday in April as S.A. Bowman's motion was approved, and predictably, an aging Ludlow was once again elected president despite his call for "a man with more youthful energy."

The following year, Benjamin Warder, Oliver Kelly and George Frey, each a prominent board and community member, accepted the critical task of investigating Ferncliff's growth potential. Warder resolved "that the cemetery grounds be enlarged, provided the same can be done without using any assets of the association." They reported their findings to the board on May 16, 1882, announcing, "We regard the purchase of the (Moffett) lands, nineteen acres, as being desirable, but find the widow (Sarah Moffett) very adverse to making any terms of sale. Time may make some changes in the situation such that the purchase may eventually be made. It will be the will of the board to keep the matter constantly in mind, that advantage may be taken of the earliest opportunity to secure these lands.

"The lands of the Bechtle heirs, twenty-five acres on the west side of the cemetery, may be had, as we are advised, at the price of $300 per acre, and we have no hesitation in recommending that this purchase be made. The property of the Bechtle heirs does not cover the entire west line of the present lines of the cemetery association. Henry Miller is the owner of about six acres of land on the north extremity of said west line, and his portion is under tillage." The property offered significant improvements: "a good, firm dwelling house, barn and other outhouses, all of which he is willing to sell at a price of $5,000.

"The purchase of this property may be largely aided; possibly the whole purchase we may secure by subscription of friends of the cemetery who are interested in making it, in size and character, what the needs of the future of our thrifty and growing city may demand."

Benjamin Warder offered the following resolution: "Resolve, that it is the sense of this meeting that the board of trustees proceed to purchase such portion of the 25 acres of the Bechtle heirs as can be bought. Also, the Miller property (should be) purchased, provided not less than $5,000 be raised by subscription toward such purchase." William Warder, originator of the rural cemetery concept that became Ferncliff Cemetery, resigned in April of 1883, with Benjamin Warder selected to fill the unexpired term. John Ludlow was re-elected president, and perhaps no one but Ludlow himself was surprised.

Ferncliff's first superintendent, John Dick.

The John H. Thomas monument, the tallest at Ferncliff at 50 feet high.

Springfield Mourns

THE DEATH OF 'FATHER FERNCLIFF'

*So when a great man dies,
For years beyond our ken,
The light he leaves
behind him lies
Upon the paths of men.*

—"Charles Sumner"
Henry Wadsworth Longfellow
(1807-1882)

Speaking at the Mount McGregor, N.Y. cottage where Ulysses S. Grant succumbed to cancer at age sixty-three in July of 1885, Benjamin Harrison surmised, "Great lives never go out. They go on." The same could have been said two years earlier at the June burial of Dr. John Ludlow, today known as "Father Ferncliff." As unassuming as the man whose memory it preserves, Ludlow's gray-granite tombstone commands no special markings, no historical notations. Its simplicity refuses to boast. There's no word of his tireless devotion, selfless accomplishments, or the high regard in which he was held. "The monument is as he was," observed the *Springfield Republic* on July 5, 1887. "Clean, white, symmetrical, majestic, always commanding admiration and respect.

"Ferncliff is, as a whole, unique. The avenue by which it is entered, as we often have said, is unequalled by any park entrance in the world. With Lagonda on the one side and the wall of fern-like (covered) rocks on the other, with fountains planted in intervals, it presents a picture one could afford to go far to see. Art has done much to supplant nature in Ferncliff."

Unbeknownst to many a cemetery visitor, this tranquil, tree-lined gem of an island, tucked amidst city roads bustling with the business of life, was largely the realization of one man's vision—the product of John Ludlow's lifelong crusade. Elected by a board of trustees he devotedly served, this man of quiet distinction was tapped on September 4, 1863 as the Springfield Cemetery Association's inaugural president, and until his death on June 10, 1883, remained its only president. Re-elected annually, despite his occasional, humble objection, Ludlow served honorably, if not remarkably for twenty years, guiding Ferncliff from debt-ridden dream to meticulously maintained reality. Just two months prior to his death at age seventy-two he'd been elected to serve again in 1884.

So prominent was Ludlow during Ferncliff's initial two decades of development that interested historians perusing the more than 140 pages of cemetery minutes experience a sense of loss when it's noted in a June 14, 1883 entry, hand-written in ink, that Benjamin Warder, chairman of the executive committee of the board of trustees, "has called a special meeting for the purpose of electing a president of the board recently made vacant by the untimely death of John Ludlow." Board members responded by unanimously selecting Warder as his replacement, and although one man's storied tenure came to an abrupt end, his dream and influence live on in the rustling trees and flowering blooms of Ferncliff.

With considerable remorse, committee chairman E.G. Dial presented the following: "Resolved, that in the death of Dr. John Ludlow, this association is called to lament the loss of its devoted president, who from its first institution, September 4, 1863, and subsequently from year to year, has been unanimously chosen to that office. Resolved, that his careful attention to the affairs of the association, his prudent council in his business management, his interest in every proper improvement and his general good taste in the ornamentation of the grounds have contributed much toward making our cemetery what it is —the most attractive feature of a public character within the city limits. Resolved, that his quiet, genial, gentlemanly intercourse with the members of the association has been highly appreciated on our path, and we take pleasure in our recollection of the kindly, reciprocal sentiments of regard which have even existed between its members and our departed friend. Resolved, that these resolutions be entered into the minutes of the association and a copy thereof be sent to his bereaved family."

Ludlow's legacy, unannounced by a modest, angular gravestone in Section B, Lot 9, nonetheless remains the epitome of Henry Wadsworth Longfellow's ageless observation in Hyperion: "He spake well who said that graves are the footprints of angels."

John Ludlow
Ludlow monument

The GAR mound in spring.

Sacrifice Honored

THE BIRTH OF SOLDIERS' MONUMENT

Rest on embalmed and sainted dead!
Dear as the blood ye gave;
No impious footstep shall here tread
The herbage of your grave;
Nor shall your glory be forgot
While fame her records keeps,
Or Honor points the hallowed spot
Where Valor proudly sleeps.

—"Bivouac of the Dead"
Theodore O'Hara (1820-1867)

Heeding John Ludlow's legacy of ongoing beautification and expansion of Ferncliff Cemetery as a blueprint, board members approached J.B. Rubsam in June and July of 1883, curious as to whether his land along the south side of Buck Creek might be for sale.

"To the board of trustees of Ferncliff Cemetery, I will sell the three most northernly lots in my edition opposite the cemetery grounds for $300, reserving an alley twelve feet off the south side of said lots," he proposed. "This offer must be accepted before council meets tonight. Otherwise, I must withdraw it, as I present my plat to the council tonight." His offer was happily accepted.

Following the September 11, 1883 resignation of board president Benjamin Warder, whose extensive travel abroad has prevented him from carrying on his responsibilities, C.H. Bacon was elected his successor. The board's focus returned to the soldier's mound. A Civil War veterans' association, the Grand Army of the Republic (GAR) appointed a committee of three to confer with the cemetery association. A suitable location formerly known as Indian Hill was agreed upon and grading began.

"Report to the committee on the Indian mound," E.G. Dial announced on October 1, 1883. "At the meeting of the stockholders held June 24, 1864, it was resolved that the board of trustees be authorized to set apart the circular mound as a burial place for deceased soldiers from Clark County, and that the same be tendered to the public without charge as a site of a soldiers' monument to be approved by the board of trustees."

An annual president's report dated November 23, 1864 added, "At the dedications of the grounds on the Fourth of July, 1864, the trustees concluded to set apart the grounds within the circle around the Indian mound as a place of burial of the soldiers of Clark County, who died in the service of the county during the (Civil War) rebellion, and as a fit place to erect a suitable monument to their memory." The Honorable Samuel Shellabarger took an opportunity during the holiday dedication to update citizens on the site's progress and the news was graciously received.

On October 18, 1883, board members met with Mitchell Post #45 of the GAR and agreed to terms. Charles W. Shewalter authored the following resolution: "Whereas, when Ferncliff Cemetery was laid out and dedicated, the cemetery association authorized the trustees of the cemetery to set apart a certain circular mound as a burial place for deceased soldiers from Clark County, and that the same be tendered to the public free of charge as the site of a soldiers' monument, and as this mound has undergone some improvement within the past year by the members of the Mitchell Post #45, GAR, we now learn for the first time that this piece of ground known as the soldiers' mound has never been deeded to Clark County by the trustees of the cemetery, and as such, no further improvements can be undertaken

The Gardiner monument, Section I.

Kelly's Lake in 1886.

before such deed can be made.

"Resolved, that we must respectfully request that a deed be given conveying this mound to the commissioners of Clark County as a burial place for deceased soldiers. We, the members of Mitchell Post #45 GAR are agreeing to use our best efforts to have a monument placed on said mound."

Upon S.A. Bowman's motion, the following resolution was subsequently, and unanimously, passed: "Resolved, that the communication of Mr. Charles W. Shewalter, in behalf of the GAR, requesting a deed of the central circle in the cemetery be referred to the members of the association at the next annual meeting, with the recommendation that the cemetery association adhere to its original action of June 27, 1864 as to said central mound, and that the clerk be directed to advise Mr. Shewalter that the trustees have no authority to make the deed requested."

On May 1, 1884, Bacon and the board entertained a proposal from Mrs. Sarah Moffett for the sale of "seven or eight acres of land (lying) between Yellow Springs Street and the east line of the cemetery at $800 per acre—$2,000 in cash and the balance in five years with eight percent interest." Although this offer was declined, Bacon was appointed a committee-of-one to issue a counter-offer of $500 per acre.

Meanwhile, drainage issues became increasingly problematic at Sylvan Hill. The board decided to investigate solutions, including the potential cost involved in draining the small lake at the foot of Sylvan Hill into a rock face in front of the Machpelah vault (where bodies awaited burial during the winter). By May 9, money was allocated and work was scheduled to begin.

By June 1, 1885, board members again turned their attention to the budding soldiers' mound, discussing the rights and privileges GAR would have to the site. E. G. Dial's resolution offered the group long-term caretaking and beautification opportunities, noting, "... whereas the association known as the Grand Army of the Republic (is) reported as desirous of grading, sodding, planting and improving said ground now, therefore, be it resolved that the (GAR) is hereby solicited and authorized to grade, plant and improve said circular mound, subject to the approval of the trustees of the association. Resolved, that the board of trustees be directed to notify the (GAR) of the action of this association in this matter."

By fall—September 16, 1885—the GAR was granted final permission, "under the direction of the superintendent, to grade and sod the central mound in the cemetery, and enlarge the road around the same five feet in width." With approval of this motion, what could have become a sensitive, ongoing issue among the county's Civil War veterans was, instead, quietly resolved.

Sylvan Hill

SYLVAN HILL

As of June 29, 1885, waterworks concerns abounded. O.S. Kelly, S.A. Bowman and C.H. Bacon authorized the engineering of a feasible, two-fold plan—one capable of solving the cemetery's ongoing drainage problems while, at the same time, bolstering water supply to maintain grass, shrubs and trees located out of range of the area's natural springs. Three-inch iron pipes were utilized and, within four months, connected to city mains.

By August 3, 1886, progress was also achieved regarding desired cemetery expansion. Sam Bowman reported that the widowed Sarah Moffett agreed to the board's "purchase of 7.26 acres of land ... for the sum of $3,000, payable one-half case and the balance on or before the first day of August, 1889, with six percent interest payable semi-annually. Kelly consummated the deal, drawing an order on the treasury for $1,500.

An October 4, 1886 board of trustees meeting took place at Kelly's Lake, located in front of Sylvan Hill. At issue? Whether to remove the bank that separates a small lake at the base of Sylvan Hill from the recent and more decorative lake recently donated and placed by its eventual namesake, O.S. Kelly. Benjamin Warder moved that sincere appreciation be "hereby heartily tendered Mr. O.S. Kelly for the efficient means in which he (has) completed the lake in the cemetery grounds, and that his proposition to unite the two lakes by taking away the dividing land meets the approval of the board, and will still further increase our sense of obligation to him." The new lake, constructed at Kelly's expense, was officially christened "Kelly Lake" on Feb. 23, 1887, and to this day, remains a most scenic and serene centerpiece atop Ferncliff Cemetery's original grounds.

Members also announced the purchase of the Moffett property—"7.26 acres of land on the east line of the cemetery—accomplished after a continuous effort of about eight years. The sum paid (is) $3,000. (Mrs. Moffett) still owns about 12 acres lying between Yellow Springs and Plum streets that the trustees are willing to purchase at a fair price, and which would be a very desirable addition to the grounds."

"Father Ludlow" would have been proud.

Tranquil Kelly's Lake, bordered by Cypress trees.

Sacrifice Honored

House on a Hill

HOUSING THE SUPERINTENDENT (PART II)

Home, in one form or another, is the great object of life.

—From "Gold Foil: Home"
Josiah G. Holland (1819-1881)

The second house built for Ferncliff superintendents.

More than 20 years passed before the board of trustees revisited the need for a new superintendent's home. H.M. Shepherd utilized an April 3, 1888 annual meeting to discuss building of a new dwelling house and office for the superintendent. Emphasizing that the present home remains unfit for occupancy "by reason of dampness caused by its proximity to the rocks," Shepherd enticed fellow board members Kelly and Morrow to join the discussion. John Thomas motioned that the matter be referred to the executive committee and requested full investigation of the subject, including workable plans and suitable cost estimates.

The April 2, 1889 meeting thus opened with a reading of a resolution that called for "a new dwelling to be built for superintendent Dick on the hill above the gate, just north of the present residence," preferably of stone or brick. "Resolved, it is the sense of this meeting that a new residence for the superintendent of Ferncliff Cemetery should be built this year," Kelly announced, "and the sum of $3,000 be set aside from the first sales of lots sold on Sylvan Hill to aid the same."

Within days, the long-talked about project finally began to take shape. Board members visited the Plum Street site just north of Dick's dwelling, discussing the exact location and type of house and any differing or opposing views. Executive committee member S.A. Bowman moved that costs be investigated and a plan be submitted.

Prominent local architect Charles Cregar was asked on April 8, 1889 to prepare a sketch of a plan and, when finished, to notify the clerk. Within a week, Cregar offered three plans, two of which were deemed too expensive. Changes and cost estimates were sought regarding his third, most feasible plan, and eventually, the group settled on a house constructed entirely of stone.

Cregar's services were officially employed on July 12, 1889, with cost of the home not to exceed $6,000. O.S. Kelly oversaw construction plans and competitive bids, and reported on July 17 that cost estimates ranged from $4,970 to $6,300. Ross & Hullinger was awarded the contract, with board members insisting, perhaps unrealistically, that the home be completed and its keys delivered by no later than January 1, 1890—a seemingly impossible five-and-a-half months. Ongoing illness within Superintendent Dick's family appeared to be hastening board members' concerns and expectations, and for peace of mind, a contractual clause was added stipulating that $25 be forfeited each month that construction surpassed the deadline.

Ironically, given the board's meticulous (and time-consuming) handling of the project—everything from design and elevations to construction materials was heavily scrutinized—handing a prominent, local construction firm just short of six months' completion time for a project of such magnitude likely encouraged precisely what cemetery officials were trying desperately to avoid: shortcuts in both quality and detail. Nevertheless, board members stuck to their demands and Ross & Hullinger accepted the challenge, agreeing to complete the work on or before Jan. 1, 1890. The firm was to be paid $4,979 in four installments, with the first payment of $1,500 due when the second-story joist was in place. A second payment of $1,100 was due upon roof completion, and a third, in the amount of $900, awarded once all plaster work was complete. A final payment was planned when the home's keys were delivered.

Clearly, board members held the upper hand in this seemingly lopsided agreement, further contracting that Ross & Hullinger be bound to pay $2,500 to Ferncliff Cemetery Association trustees (or complete the work at builders' expense) should it fail to meet any condition of the agreed-upon contract.

But eventually—and expectedly—the project didn't flow according to plan. During a January 24, 1890 meeting it was reported that Ross & Hullinger had failed to meet their deadlines. January 30 brought the following statement: "To the Cemetery Association of Springfield, Ohio, we the undersigned bondsmen of Ross & Hullinger hereby notify you that we will complete the building as per the contract entered into between ... Ross & Hullinger and the Cemetery Association, as we are permitted to do under the conditions of our bond, and we hereby notify you that we have, this day, entered into an agreement with A.B. Smith to complete said contract for the sum of $1,949, and authorize you to pay—A.B. Smith said amount, as the same may become under the contract made between you and Ross & Hullinger."

On April 1, 1890 the board likewise resolved "that the penalty imposed upon the contractors of the new building at the cemetery, for the failure to finish the same at times specified in the contract, should not be exacted provided that all claims are paid by them, their assigners or successors." Though a bit awkward and stressful, issues concerning a new home for John Dick were finally resolved.

Today, Ross & Hullinger's magnificent achievement sits stoically north atop the limestone cliffs, anchoring the Plum Street entrance and serving well those superintendents who've followed.

Historic image of superintendent's house shortly after completion, 1890.

The Leaning Rock along the Plum Street entrance road.

A Wise Economy

WATERWORKS BEAUTIFY THE GROUNDS

*Gone are the living,
but the dead remain,
And not neglected; for
a hand unseen,
Scattering its bounty like
a summer rain,
Still keeps their graves and
their remembrance green.*

—"The Jewish Cemetery at Newport" Henry Wadsworth Longfellow (1807-1882)

Understanding Ferncliff Cemetery's evolution is aided little by previously written works; history's only clues lay preserved in handwritten meeting minutes left behind by the concerned citizens who created it.

Their notes from September 20, 1894 reveal that sewer construction on the grounds had become a priority. S.S. Taylor was awarded the job on the strength of a modest $4,200 bid. Always an issue, lack of funding prompted some board members, including S.A. Bowman, to protest sewer construction, but the naysayers were eventually overruled. Ironically, the project eventually ran $150 over budget.

"The only important work in the way of improvement (is) a commencement of the building of a large sewer through the east end of the cemetery grounds and running from McCreight Ave. on the north to the main entrance avenue on the south," reads an April 3, 1895 report. "It (is) a work that (seems) to be needed, and its completion (will) bring into market a large strip of land that, without a work of its kind, (can) not be used for burial purposes. Costs of it may be large, but the money investment (is) thought (to be) a good one and will bring a good interest.

"The work (is being) pushed to completion and, when done, (will) give a large number of desirable and valuable lots. The income from the sale of lots will fall short of the previous year, as shown by the reports given during the annual meeting."

Today, the results of this once-debated sewer project ensure year-round beauty along Ferncliff's Plum Street entrance, sustaining the picturesque waterfalls that serve as the centerpiece of many a photographer's composition. A mural in the foyer of the administration building both captures and preserves their beauty.

Running below Plum Street and emptying into Buck Creek, the falls later became part of a more expanded waterworks project that tapped into the city's sewer system.

With drainage and water supply problems behind them, the Cemetery Association turned its attention to securing additional land for the benefit of future generations. On September 15, 1895, Mr. H.G. Bosart offered to the association an attractive piece of land—part of the Moffett property. Gustavus S. Foos, association president, reported in November his purchase of the Moffett land for $650. Without hesitation, the board approved and issued a check for the agreed-upon amount.

Just two months later, the board mourned the death of S.A. Bowman, "a devoted friend for the improvement of the cemetery." In accordance with cemetery bylaws, an election was held January 8, 1896 to fill his vacancy.

MODERNIZATION CONTINUES

Foos, meanwhile, reported at an annual meeting in April of 1897: "Nature has been beautiful in supplying us with a beautiful location. Yet man's ingenuity can so improve the spot as to make it almost marvelous, and we should all, as citizens, feel proud of being possessed of such a burial place. There remains yet much to be done to

Ferncliff's picturesque waterfalls.

complete the work thus begun, but what it will be when completed can be seen by a look at Section M, as now finished. In our judgment, it should be the aim of the association to instruct or authorize the trustees to acquire by purchase, at as low a price as possible, all the grounds lying between the east line of the cemetery grounds and Plum Street."

Foos added, "Since our last annual meeting, a large amount of work has been done at the cemetery grounds in the way of filling, grading and putting in shape the large amount of grounds on the east side of the cemetery that, since its inception, has been a wasteless piece of ground —an eyesore to those interested in bringing this cemetery to (what) all would have it to be. Instead of an open waterway that wound itself in a torturous way all over that part of the grounds, and in times of (rain) washing away everything in its course, a large, new arch sewer has been built from north to south through the grounds, thus confining its water to a regular course and reclaiming ground that, in time, will bring a revenue to the association of many thousands of dollars. While this has been done at what would seem considerable expense, we regard it as a wise economy. The improvement thus undertaken and as far as completed has been largely the work of the superintendent's direction. This work was decided upon by the trustees only after many thorough deliberations and interchanges with views of the board."

On April 8, 1898, board president Gustavus Foos encouraged fellow trustees to review Ferncliff's constitution, with an eye toward change. The board resolved to amend Article 2, allowing the association to meet "annually on the Tuesday following the first Monday in April of each year." Further, "Nine members of the association shall constitute a quorum." Article 6 was likewise modified, signaling that "the annual election of trustees and clerk shall (now) be held on (the) Tuesday following the first Monday in April of each year."

Board member E.O. Bowman, who has replaced recently deceased S.A. Bowman, moved to add Article 14, which reads, "That 20 percent of all monies received by this association from the sale of lots in its cemetery shall be set aside as a permanent fund for the future care and maintenance of said cemetery grounds, and the same shall be invested on first mortgage or other security under the direction and control of the board of trustees, and that no part of the principle or interest thereof shall be expended for any purpose except for a resolution of the association to be passed at an annual or called meeting thereof."

Motions regarding the constitution's structure were also passed, further modernizing the document to allow the board to function more effectively. Industrialist O.S. Kelly was elected association president, succeeding G.S. Foos. One of Kelly's first acts of presidential business was to request the establishment of a committee to research and compile a written history of Ferncliff for presentation in pamphlet form. J.H. Rodgers was appointed a committee-of-one to complete the project, gathering information from past minutes to author a brief, historical sketch of the grounds suitable for publishing in accordance with the cemetery's constitution and bylaws.

Moral Purpose

BURYING THE INDIGENT

If a free society cannot help the many who are poor, it cannot save the few who are rich.

—Oct. 26, 1963 Address at Amherst College, President John F. Kennedy (1917-1963)

Once again concerned with expansion, O.S. Kelly and the board established a committee on December 16, 1901 to confer with the Board of Public Affairs regarding a twenty-acre parcel of land west of Ferncliff Cemetery and owned by the city. Trustees entertained the idea of having the ground transferred to the association under various conditions and restrictions. A portion of the ground was eventually set aside for use by the city as a burial place for Springfield's poor, with its size not to exceed five acres and the association charged with its maintenance.

Strangely, the board withdrew its proposal just four months later, on April 8, 1902, with nary a reason given for its change of heart. Perhaps encouraged by the prospect of obtaining adjoining land offered and owned by a woman referred to only as "Mrs. Leightly," board president C.H. Pierce extended a $500 offer, but Leightly balked at the deal come September. Frustrated that cemetery expansion was once again thwarted, a board concerned with burying the indigent resolved on February 7, 1903 to furnish Springfield with 800 graves for interment of the poor, without expense to the cemetery, providing that the city convey a 20-acre tract of land lying west of Ferncliff and the road connecting the two properties.

Interestingly, the board also moved during this same February meeting that association rules and regulations be amended to read: "On or after March 1, 1903, no interments shall be permitted in the cemetery on Sunday(s), except a special permit be obtained ... from the executive committee in each case, and that such special permits shall only be granted by them in good cause."

A March 13, 1903 special meeting resulted in a new proposal to the BPA: the city would transfer to the cemetery association the desired twenty-acre tract of land in exchange for the association furnishing the city with 800 single graves for burial of the city's destitute poor. Interments were to be completed within fifty years from the land transfer date, with the graves under the exclusive care and control of the cemetery, and as such, subject to its rules and regulations.

A July 24, 1903 local newspaper headline announced the agreement: "Ferncliff's Grounds Enlarged for Dead of Future Years." The cemetery association now controlled 185 acres. "Two deeds were approved by the association at its annual meeting held last evening in the city building," the article stated. "One (is) for 19 acres of ground in the Oakwood Cemetery, transferred by the city, and another (is) for a tract of 14 acres of ground sold to the association by James A. McNally, executor, for $2,500. The city gets 800 graves for burial of its poor and additional at $5.00 apiece, and the association takes care of the entire property the same as it does Ferncliff Cemetery. The association (begins) upon the improvement of the driveway into the cemetery about the first of August, just as soon as the material can be obtained."

DEATHS OF O.S. KELLY AND ASA S. BUSHNELL

E.O. Bowman reported with considerable sadness at an annual meeting of April 12, 1904 that a resolution of respect be extended to late president O.S. Kelly and his family. Among Springfield's most prominent of men, the famed industrialist passed into eternity buoyed by a tribute preserved for the ages.

"As members of the society," Bowman read, "we deplore the recent death of our friend and fellow member, Oliver S. Kelly, and desire to spread upon the minutes a memorial of our affection and esteem for him as a member of the society—and as a friend and as a man. Therefore, be it resolved that (to) the devotion of Mr. Kelly, both as a member of the society and, for many years, as president, to the eternal of Ferncliff, the gratitude of the members of the society and of those of our citizens whose dear ones have found a lasting resting place in the spot which his loving has done so much—that his conduct in all walks of life has been that of a brave, generous, public-spirited gentleman, and that we shall always cherish in grateful remembrance his uniform kindness and consideration during the many years we have been associated together."

The association mourned a second time in April, meeting this time to replace the late Asa Smith Bushnell on its twenty-five member panel. Bushnell, a Republican who served as Ohio governor from 1896-1900, died January 15, 1904 from complications of apoplexy (an early term for "stroke"), three days after being stricken aboard a train while returning from inauguration ceremonies in Columbus for Ohio's newly elected governor, Myron Timothy Herrick.

By spring, his passing had created a significant void not only among board members, but within the city as well. A prominent and highly regarded Ohioan, Bushnell once served as president of Warder, Bushnell and Glessner, an agricultural machinery company which later merged with other companies during the early 1900s to form International Harvester, today known as International Truck & Engine Corporation. He also served as president of both First National Bank of Springfield and Springfield Gas Company.

Springfield mourned as another great man of industry and community was gone.

MOVING ON

Thoroughly devoted to making Ferncliff Cemetery a place for the living, the board of trustees met on November 25, 1904 for the purpose of "consulting and advising in regards to building a chapel and entrance gates to the cemetery." To assure proper funding of this most significant and symbolic structure, the board resolved that "10 percent of the gross sales of lots in Ferncliff Cemetery be set apart as a special fund known as the Chapel and Entrance Fund, which together with any contributions that may be made thereto by lot owners in the cemetery shall be invested separately, and the interest accumulated until such time as in the judgment of the

The Bushnell Mausoleum, Section P

board of trustees is sufficient for the purpose of erecting a chapel and suitable entrance gates at the main entrance of the cemetery on Plum Street."

Soon, though, members' attention turned once again to the passing of yet another Ferncliff "mover and shaker"—a key player in Springfield's historic rural cemetery movement who proudly and quietly developed and groomed Ferncliff's once-empty grounds into a fragrant, colorful town centerpiece—a place of grandeur in the heart of industrial-based Springfield.

Newly elected board president C.H. Pierce called to order a special meeting to mourn the passing of Ferncliff superintendent John Dick, a man whose death stirred a complicated mixture of sorrow and concern among trustees.

"The death of such a citizen as the late Mr. John Dick is no ordinary event in any community," Pierce said, reading aloud the proclamation he'd authored. "He was for more than 43 years the first and only superintendent of our beautiful Ferncliff, a cemetery surpassed and, in some ways, equaled by none other in our land. It is due to his memory to say what it is, as seen by us, and what its claim in the eyes of an admiring world elsewhere is, due largely to his taste and skill, and his name will be associated with it in our minds and hearts, and also in those of our children's children.

"So long a service, and so valuable, made him known to and esteemed by our people, generally as is the good fortune of but a few men. As a man he compelled respect by his courtesy of manner and kindness of heart. As an official he was a model of faithfulness and duty, and he was readily equipped to the work to which he devoted his long and useful life. His personal relations with his board were the pleasantest. Always cooperating with us promptly, heartily and intelligently, and always making for the improvement of our city of the dead. We gratefully pay this tribute to his work and direct that it be entered on the permanent record of the board."

Fittingly, James Dick was appointed to succeed his father, a humble man who, by serving so admirably since Ferncliff's inception in 1863, achieved greatness in the eyes of his family, his peers and a grateful community.

As John Dick would want, addition to Ferncliff's grounds continued despite his loss. In April of 1907 cemetery president J.W. Stafford contacted Benjamin F. Prince regarding the purchase of a fourteen-acre tract adjoining Ferncliff on the west. The desired sum was $2,000 upon condition that Prince assists in the work of vacating the street that runs along the property's north side. By May 14, Prince agreed to sell the tract for the agreed-upon price, and paid upon delivery of a sufficient general-warranty deed.

Though the specific date isn't clear, it's likely in the fall of 1907 that the board returned to its vision of constructing a chapel on cemetery grounds, hiring Guy Lowell of Boston, Massachusetts to prepare plans and specifications in accordance with previously submitted sketches.

Throughout Ferncliff's expansion, the cemetery association appeared quite cooperative with the city of Springfield, as indicated during an annual meeting on April 14, 1908, when members passed a resolution stating that the association "will gladly work with the city" in granting it the rights to the land belonging to the association, which lies south of Buck Creek, so that the proposed park extension plans can be completed.

In an ongoing quest to work together for the betterment of the Springfield community, the farmland that was granted to the city in 1895 by John and David Snyder—today known as Snyder Park—received additional acreage due to the benevolence of the cemetery association.

The Gladfelter Mausoleum, Section Q

Sacred Architecture

RAISING FERNCLIFF CHAPEL

In the midst of the grounds stands a very beautiful chapel, rearing its graceful form above that long row of reverend elms. It is one of the most interesting and appropriate of the class of structures adapted solely for burial purposes; and embosomed among the ancient and stately trees, the whole forms a picture of nature and art combined, not easily to be surpassed.

—"Abney Park Cemetery: A Complete Descriptive Guide, 1869" Rev. Thomas Baker

Historic image of Ferncliff Chapel.

During the early 1900s, Ferncliff Cemetery continued to expand as Springfield visionaries planned for both construction of an on-site chapel and beautification of the Plum Street entrance. The board of trustees hired Guy Rodgers, a noted architect from Boston, Massachusetts, to prepare plans and specifications. Other architects, too, were encouraged to submit bids, but association president John Bushnell favored Bostonian Guy Lowell. Lowell met with the board at the cemetery gates on August 13, 1908 to select a suitable chapel site, then contracted with prolific local builder William Poole whose many projects included the Frank Lloyd Wright designed Burton J. Westcott house on East High Street. Poole submitted an initial cost estimate of $13,140.50.

Upon close examination of the chapel's foundation, board members identified a dampness problem with the original site selection and asked Lowell to come at once to advise them on possible relocation plans. On October 10, 1908, buoyed by Lowell's expert advice, members moved to relocate construction of the chapel's foundation. Despite the delay, the board's eagerly anticipated chapel was nonetheless taking shape.

Once again concerned with expansion, the board rescinded an offer to purchase the Moffett land bordering Plum Street. Instead, a screen of fifty elm and evergreen trees was planted along the east line, sufficient to hide an eastward view. In April of 1909, board members returned their attention to chapel-building, agreeing to relocate the site above the cliff line, west of the main entrance and north of the Machpelah vault. The previous foundation was destroyed and the site returned to its original grassy appearance.

By March 22, 1911, the board of trustees was meeting inside the new chapel, complete with furniture and accessories. Members proudly inspected the magnificent building, offering a smattering of desired modifications that Lowell agreed to address at his earliest convenience. A well-built chapel was finally completed—and with full approval from the association. For a while, the site housed board meetings as members shifted from the downtown law offices of Bowman & Bowman.

Within a month, the Daughters of the American Revolution were authorized, under newly elected president John Hoppes, to erect a memorial tablet on the cemetery grounds, and design plans and suitable locations were requested for review at a later date.

By spring of 1913, the association, headed by President Charles L. Bauer, realized that rules governing the operation of motor vehicles inside cemetery grounds had become a necessity. In the interest of landscape preservation, the board's adopted motion mandated that automobiles may enter at any time during the day "from the north gate only and proceed as far as the Soldiers' Mound only. Lot owners may enter from the north gate only from the hours of 7 and 10 a.m. and proceed to their respective lots." The move toward modernization was substantial given that automobiles had not previously been allowed on the grounds due to damage that vehicle weight can cause to cemetery roadways. Horse-drawn carriages had created similar concern. However, lot owners began to request a change in policy so that they might more easily access and tend to loved ones' burial sites, and public demand forced the board's hand.

EXPANSION AND GROWTH

On May 9, 1913 the Daughters of the American Revolution met with the executive committee to discuss placement of the memorial boulder and plaque furnished by the DAR in lasting tribute to soldiers who fought for our independence from Great Britain. They selected a suitable location for the stone, complete with memorial plaque, across from the newly constructed chapel.

The board of trustees, meanwhile, revisited the ongoing issue of land located along the south side of Buck Creek—land to be deeded to the city of Springfield for park purposes—during its annual meeting of April 14, 1914. Re-elected president Bauer presided.

The Moffett land issue resurfaced as board members passed a motion empowering the purchase of two tracts of property owned by Harvey Moffett and his sister at a price of $1,500.

Ferncliff employees in 1891

Land was a premium, and the board took advantage of every opportunity to enlarge the cemetery's grounds. The Moffett lands remained long-standing issue of debate for the board. Members now had an opportunity to purchase more lands, as indicated by notes taken at the April 18, 1915 meeting. Two tracts of land located west of the cemetery—a 7.32-acre Miller family property and a six-acre Harrison tract—became available, and the board urged Bauer to proceed with purchase, not to exceed $7,500 for the Miller property and $3,000 for the Harrison land. Bauer finally brought closure to the issue in June with a purchase fully supported by the board.

An interesting side story developed during the year with the board passing a motion that automobile funerals were now allowed to enter either cemetery gate at any time during the day, according to a May 25, 1915 report. Concern about the entrance of automobiles, however, remained an ever-present challenge to Ferncliff workers as they labored to properly maintain its roadways.

Not long after the Moffett purchase was sealed, more land became available. An agent representing William Miller offered to sell the association a 25.54-acre tract north of McCreight Road for the sum of $15,000. Vice president E.O. Bowman authorized the treasurer to pay the asking price and Ferncliff Cemetery expanded once again.

WORKERS UNITE

No association, though, functions without its share of occasional problems, as evidenced by events in April of 1916. Meeting notes reveal that a difference of opinion emerged between Superintendent James Dick and the board of trustees regarding a strike among Ferncliff workers. As a result of being unable to reconcile their disagreement over the issue, Superintendent Dick felt the best course of action for all concerned was to tender his resignation. During the annual meeting of April 11, Charles L. Bauer stood before the association and read aloud James Dick's resignation letter, which the executive committee graciously accepted without dissent. His resignation marked the end of an era at Ferncliff. For over half a century a member of the Dick family, first his father John, then later James, lovingly tended Ferncliff as it grew from simply a splendid concept into an enduring legacy. Acting quickly to fill the void, the association selected Stanford J. Perrott as Ferncliff's newest superintendent.

In April of 1918 the association altered Article 2 of the cemetery's constitution to move annual meetings from the second Tuesday in April to the second Tuesday in May. Edmund O. Bowman was elected the association's next president. By May, the hazards associated with motor vehicle use of the cemetery surfaced once again as an issue of concern for the board, which resolved that "owing to the imminent danger of collision between motor vehicles and—horse vehicles on the steep roads in the cemetery grounds, which might result in loss of life or serious injury to the people using the roads—and also to protect the roads from excessive wear of motor vehicles climbing up and down on account of the steepness of the grade—it (is) necessary to completely change the method of using vehicles in the cemetery grounds.

"It is therefore ordered that on and after the 27th day of May, all vehicles entering the cemetery shall do so through the McCreight Avenue entrance gate, and that from and after that date, the Plum Street entrance shall be used only as an exit from the cemetery grounds. In order to carry this into effect, the superintendent is instructed to place a watchman at the Plum Street entrance to prevent any vehicles from entering. And it is further ordered that the clerk be instructed to give notice to all the undertakers, in writing, and also to give public notice, by publication in the newspapers, to lot owners and other persons visiting the cemetery of this order."

In December, a new entrance running from St. Paris Pike through the ravine north of present-day McCreight Avenue was proposed, along with a preliminary plan for construction of a new administration building. Both actions were part of the board's ongoing effort to improve a tract of cemetery land located along McCreight Avenue that held Ferncliff's original administrative office.

The Mast and Blee Mausoleums, Section O.

Garden of Stone

THE DAYTON CONNECTION

*Lay me down beneaf de
willers in de grass,
Whah de branch go
a-singin' as it pass;
An' w'en I's a-layin' low,
I kin hyeah it as it go,
Singin', 'Sleep, my honey;
tek yo res' at las'.*

—Inscription on the
Woodland Cemetery grave
of poet Paul Lawrence
Dunbar (1872-1906), from
his dialect poem,
"A Death Song."

The Spanish-American—WWI Soldier's Mound

With E. O. Bowman succeeding himself as president, the association made a meaningful and historically significant decision in May, 1919 to send its board of trustees to Dayton, Ohio's Woodland Cemetery. Interested in investigating the management and operation of the nearby property, the association hoped to bring new ideas to the operation of Ferncliff Cemetery.

Established in 1843, Woodland today boasts 200 acres, some 100,000 monuments, and more than 3,000 trees, and contains some of the country's most beautiful mourning figures, angels, urns and mausoleums. The Wright Brothers are buried there, as is humorist Erma Bombeck. Poet Paul Lawrence Dunbar is laid to rest there, too, along with paper magnate George Mead and former Ohio governor and presidential candidate James M. Cox who, alongside running mate Franklin D. Roosevelt, lost the 1920 presidential race to Warren G. Harding and Calvin Coolidge. Other Woodland notables include John H. Balsley, inventor of the folding step-ladder, Loren M. Berry of Yellow Pages fame, and entrepreneurs John H. Patterson (National Cash Register), George Huffman (Huffy Bicycles) and John Glossinger, creator of the Oh Henry! candy bar.

One hundred and sixty-five specimens of native Midwestern trees grace Woodland's shaded and gently rolling hills, many more than a century old. Its Romanesque gateway, chapel and office, completed in 1889, are listed on the National Register of Historic Places, with the chapel boasting one of America's oldest Tiffany windows. During the 1913 flood, Dayton's highest point, ironically, became a life-saving place of refuge. Featuring one of the area's finest arboretums, Woodland remains anchored by its rock- and bronze-faced mausoleum featuring twenty-two varieties of imported marble as well as twelve stained glass windows inspired by great literary works.

But Woodland's early importance to Ferncliff officials was its billing as one of the oldest "garden cemeteries" in the U.S. A new concept in the nineteenth century, the rural-cemetery approach provided both a beautiful and hygienic setting for burials to take place away from the densely populated city of Dayton.

Founded in 1841 by trustee John Whitten Van Cleve, Dayton's first-born male child, Woodland began with forty acres and a vision: cemeteries should be beautiful, comforting places for the living to visit the dead. The site was thus named "Woodland" for its natural, wooded charm and expansive hilltop views, and today remains among the country's five oldest rural-movement cemeteries.

At the time of Ferncliff officials' visit, the Woodland burial of famed aviator and inventor Wilbur Wright was just six years old. Trustees were wrestling with issues of concern and expansion similar to those at Ferncliff: fashioning rules to regulate cars and automobile hearses within the cemetery while raising a shelter house, chapel, stone vaults and iron gates, and constructing a pumping station to deliver water to the grounds' summit.

Ferncliff officials were understandably intrigued.

OTHER BUSINESS

In July of 1919, Bowman called for a motion to once again change the annual meeting date, this time to the "last Tuesday of January of each year." Nine members shall constitute a quorum, with the fiscal year ending on December 31. These relatively minor issues passed unanimously.

Of more importance was an old school house—one located across the street from the McCreight Avenue entrance that officials had converted into an administrative office. The renovation served the association well for several years, but in September of 1918 the board of trustees considered vacating the building. By February, the association waived all claims to the building and any subsequent damage resulting from its evacuation—an action that followed in line with the group's motion to improve the tract of land upon which Ferncliff's old office was located in preparation for the new entrance off McCreight Avenue.

The Soldiers' Relief Commission of Clark County approached the association in December of 1918 and requested that a portion of the grounds be sold to the county and designated 'Section R' for a proposed soldiers' plat at a price of 75 cents per square foot. By March of the following year, sale of the Clark County War Chest of the soldiers' plat, Section R, was completed for the sum of $22,500. Money received was invested in U.S. Treasury certificates. The area was platted into lots by February, and is today known as the Spanish-American/World War I mound, highlighted by a magnificent hilltop cross surrounded by cannons of the period.

Of concern was that E.O. Bowman didn't attend the January 25, 1921 annual meeting. Vice president John L. Bushnell presided in his place, and despite Bowman's worrisome absence, he was elected to succeed himself as president—a strong stamp of approval regarding his likely leadership abilities. Less than a month later, Bowman's death was announced by Bushnell, who with great sadness replaced him as association president.

Bowman's memorial resolution reads, "Whereas Mr. Edmund O. Bowman has passed from our midst, therefore be it unanimously resolved that in the loss of Mr. Bowman, this association has been deprived of one of its most faithful and useful members, an esteemed associate and friend, Mr. Bowman was a member of the board of trustees of this association for nearly 26 years, during which time he served for three terms of one year each as president, and was reelected for another term just before his death.

"Ferncliff was always among Mr. Bowman's interests, and he gave his services to it with a love and devotion that was always most helpful and inspiring to his associates on the board. He was a member of the executive committee most of the time since he became a member of the board, and as such did much to make Ferncliff one of the most beautiful cemeteries in the state. While this board has sustained a most serious loss, we realize that his immediate family has suffered a much greater one, and to them we would like to extend our earnest sympathy and condolence.

"Resolved, that a copy of these resolutions be sent to Mrs. Bowman, and that a copy be spread upon the minutes of the association."

Once mourning had passed, association members returned to business by focusing their attention on the strict rules and regulations governing lot owners' rights when building mausoleums. The McGilvray family sought approval of a mausoleum design planned for the family that had donated land for a new YMCA—a tract on the southeast corner of North and Limestone streets. By July 22, 1922 the design plan was accepted and accompanied by a $25,000 endowment for its construction. However, the association maintained complete control over its building in order to preserve the beauty and integrity of Ferncliff's grounds.

The year 1923 brought consistency to the function of the board as it made important decisions. Motions regarding cemetery improvements gained unanimous approval. Whether buying property, creating walkways between burial sites or paving entrance roads—whether purchasing pipe fittings for proper lower-level drainage, improving Sylvan Hill (Section O), or seeding and sodding Ferncliff's borders—the board remained in agreement. Leadership stability continued to benefit the association, as did its goal to focus strictly upon restoring and improving cemetery grounds.

J. Warren Keifer

A GENERAL'S LAST WISH

*Not for fame or reward / Not for place or for rank
Not lured by ambition / Or goaded by necessity
But in simple / Obedience to duty
As they understood it / These men suffered all
Sacrificed all / Dared all—and died.*

—Inscription on the Confederate Memorial, Arlington National Cemetery.

Ferncliff Cemetery achieved national prominence during an annual meeting on January 26, 1926, when highly regarded Joseph Warren Keifer, an American Civil War general of renown reputation who also served in the Spanish-American War, requested that, although entitled to proper burial with pomp and circumstance in revered Arlington National Cemetery, he wished to be buried in "beautiful Ferncliff with my parents, wife and children." At the time, General Keifer was approaching ninety years of age, yet still held several prominent positions within the Springfield community. To have requested burial in Ferncliff Cemetery remains a remarkable testament to the character of both the man and his beloved Springfield community.

Born in Bethel Township on January 30, 1836, Keifer ranked among America's longest-living politicians when, at age ninety-six, he passed away in Springfield on April 22, 1932. He was both a brigadier general in the Union army during the Civil War and a prominent Ohio Republican, serving in the U.S. House of Representatives from 1877-1885 and from 1905-1911. A former longtime speaker of the house (1881-1883), member of the Ohio State Senate (1868-1869), and onetime delegate to the Republican National Convention (1876), Keifer established a Springfield law practice on January 12, 1858, to which he later returned in 1873 and remained in practice until his death. His union service included stints with the 3rd Ohio Infantry, the 110th Ohio Infantry (colonel), and the Eastern Theater, and action in the Battle of the Wilderness, Overland Campaign, the Valley of the Campaigns of 1864 and the Siege of Petersburg. Yet he's perhaps most well-known for capturing Major General George W. Custis Lee, Robert E. Lee's eldest son, at Sayler's Creek, Virginia in April of 1865.

Also a Brevet Major General of Volunteers in the Spanish-American War, Keifer later served as a trustee of his alma mater, Antioch College, and as president of Springfield's Lagonda National Bank. Fittingly, his papers are preserved in the manuscript reading room of the United States Library of Congress.

GOING PUBLIC

In 1927, just two years prior to the onset of the Great Depression, the association's executive committee for the first time published in the daily newspaper portions of an annual report to better inform the public of the true conditions existing at Ferncliff. This action was taken largely to offset growing sales talk from Rose Hill Burial Park agents and the possible effects that investment cemeteries could have upon older, more established cemeteries like Ferncliff. The incident appeared to be the first of an advertising nature for the association as it was forced to embrace a more aggressive campaign in order to sell lots to Springfield residents.

By April of 1928 board members revisited the idea of building a comfort station. Although formal approval for such projects seemed to always require considerable time, estimates and plans were finally approved, with the structure's maximum cost not to exceed $4,500. Local contractor A.G. Samuelson offered a $5,067.80 bid, plus architect fees, to construct both a comfort station and a shelter house. The offer was unanimously accepted.

As with any organization, governing rules and regulations often change, and the Cemetery Association was certainly no exception. Superintendent Perrott announced during a January 29, 1929 meeting that the Ohio Association of Cemetery Superintendents was sponsoring a bill before the state legislature that states, among other requirements, that 25 percent of all cemetery lot sales were to be placed into a perpetual care fund. Because the cemetery association was only putting away 20 percent, this startling announcement prompted board members to increase lot prices sufficiently enough to raise their endowment to the soon-to-be-required 25 percent.

With financial matters adjusted and improved, the board met on October 14, 1929 to discuss plans for a new administration building. Its need had been apparent for several years, and superintendent Perrott was authorized to consult with architect W.K. Shilling for the building of new administrative offices at a cost not to exceed $40,000. The new building was to be located near the present office. Shilling was well-known for the initial restoration of nearby Pennsylvania House, and would go on to build the Springfield post office, which shares striking similarities with Ferncliff's administration building. Interestingly, when Shilling passed away in 1939 he was interred at Ferncliff in Section A, and his grave was marked by a headstone that bears the influence of the Art Deco style he favored and utilized for the post office and the administration building.

By January, Shilling presented several tentative plans, which were eventually perfected for final submission. Financing became a concern, given that all available funds belonged to the permanent fund for future care and maintenance. Board members located and studied a statement in Article 11 of the constitution, under the heading "Clause of Future Care and Maintenance," which read that development that adds value to the cemetery, such as new buildings, fences, sewers or roads, or grating or platting of new sections, also provides for general care of the cemetery and can, therefore, be considered for funding. The board passed a motion empowering legal council to draft a resolution that embodied the right to "apply (monies) now on hand and which may hereafter accrue to such funds to the construction of a new administration building at a cost not to exceed $50,000."

Perrott and Samuelson secured enough Greenfield limestone for the new building, thus matching the character of material used for the gateways. The executive committee agreed to pay Shilling a 4 percent commission on construction costs for his plans and specifications, which included his partial supervision during the project.

By April of 1930, A.G. Samuelson's bid was $35,933, based on usage of Indiana Oolithic limestone as the material of choice. Samuelson was given the following conditions: 1) that he be asked to provide a bond for project completion and 2) if Indiana Oolithic limestone costs less than $5,100, the amount

The Gothic-style Bookwalter Mausoleum, Section I.

be deducted from the contracted price and 3) that the building be completed by October 15, 1930. Once again, the association instructed the committee to place time constraints upon the builder, seemingly its preferred way of business for major construction projects. A $35,933 bond was secured and split between Samuelson and the Cemetery Association.

Wallace & Co. administered tornado and fire insurance to cover the new building, and Hawkins Electric Co., the lowest bidder at $318, was awarded its electrical responsibilities. Baxter & Naftz earned the heating and plumbing contract, with specifications for a capital boiler, without stoker, from Taggart Coal and Supply Co. for $4,242. A stoker was purchased from Taggart Coal and Supply Co. at a cost not to exceed $475, and money was approved to purchase two Grecian urns for placement in front of the new building. Shilling was then authorized to complete all furnishings, including furniture, rugs, draperies, lighting fixtures and wall decor, at a cost not to exceed $3,500. People's Outfitting Co. of Springfield and Egelhoff Studios of Columbus were placed in charge of interior design.

In January of 1931 the building was insured for $37,500 and its contents for an additional $8,000. Dedication day was set for Sunday, May 24.

The Administration Building.

Widow's Walk.

Fade to Black

MANAGING PROFITABILITY

A cemetery may be regarded essentially as a museum and archives in itself.
Each gravestone is an irreplaceable and unique artifact that provides historical cultural information inherent to an individual and (his) community. Unlike most documented histories, graveyards recognize one commonality amongst everyone, and signify the past lives and existence of all. The durability of the materials used to create these monuments indicates the importance placed upon preserving the memory of loved ones.

—From "MANL Note," a publication of the Museum Association of Newfoundland and Labrador

Ferncliff Cemetery

By 1933, American's Great Depression was in full swing and lot sales at Ferncliff had dwindled. In a letter dated February 17 of that year, superintendent reports to the clerk and treasurer revealed a shortfall in 1932 lot sales receipts "to the extent that we did not save the required 20 percent for maintenance as required by the constitution and bylaws under which the cemetery has always operated." The oversight totaled approximately $1,400. Because he expected no increase in lot sales during poor economic conditions, Perrott suggested three items to consider to solve the cash-flow problem. Regarding salaries, he wrote, "all voluntarily accept reductions as follows:" superintendent: $400, assistant superintendent: $200; the clerk: $200, and another unnamed employee: $180, totaling $1,080. Additional reductions, he insisted, could push the savings to $2,000. "With that objective in view," Perrott added, "we, your employees, are making this offer wholeheartedly."

Further changes took place as association president H.C. Downing resigned to accept a position with a Nevada-based company. With "great regret," the board accepted Downing's resignation and decided, at least temporarily, not to elect a new president. Existing vice president J.B. Cartmell agreed to serve in the capacity of president until the association took further action.

Addressing Ferncliff's financial shortfall, board members composed and mailed letters to all citizens harboring delinquent account mortgages held by the association, utilizing a bit of friendly persuasion to encourage settlement in "an agreeable manner." Such measures became necessary, they explained, in order to honor their fiscal responsibility to lot owners.

Perrott subsequently noticed during late January of the following year that labor charges for grave digging were considerably below the Ohio average. While researching and corresponding with roughly sixty other cemetery officials throughout the state, Perrott also discovered they'd been undercharging in other areas as well, one example being cement vaults. Until the situation could be studied and remedied, the association motioned to enable its board of trustees to apply $15,000 in permanent care funds to maintain Ferncliff's grounds in the event that receipts remained insufficient to cover necessary expenses. Cartmell, meanwhile, officially accepted the board's nomination and was elected association president.

Concern over the cemetery's insufficient income lingered well into March of 1934. Due to delinquency of lot sales, lagging interest collections, and an inability to reduce book accounts during the past year, maintenance staff were now unable to begin timely improvements. Once again, the issue of payroll reductions was placed before the board. Members voted to reduce payroll by $2,800, signaling that Perrott and his staff would, indeed, receive salary reductions of $100 each, retroactive to January 1. To quell ongoing worries, they discussed both establishment of endowments and reduced lot rates in exchange for cash paid at the time of sale. Implementation of these and other changes was eventually rewarded. By July, income exceeded disbursements by $1,675.82.

Trustees' newfound, fiscally conservative mindset prompted them to reconsider real estate owned by the cemetery—possibly selling some of the land for its purchase price. Given the advent of a new investment cemetery (Glen Haven Memorial Park located along Route 40, roughly eight miles from west Springfield), members also questioned, again, whether an advertising push was necessary, wondering aloud whether a photo leaflet capturing the magnificence of Ferncliff might spark dwindling lot sales. They also considered employing a commissioned lot salesman. Both ideas were approved unanimously, and their implementation was to begin in October of 1934. Howard Weber, a local commercial photographer, lent his requested, professional input to what would become the booklet's overall look and design. Regarding commissioned sales, the board considered offering any new sales employee 10 percent of lot price sold, coupled with a five-percent commission if buyers paid cash upfront and in full. Members remained confident that such additions could help solve the cemetery association's more immediate cash-flow challenges.

A representative of Winwood & Company became a commissioned sales agent for Ferncliff when, on November 12, the board voted unanimously to delegate lot sales to an outside business. In return, executive committee members retained full power to act upon any problem that might arise regarding such sales, including all terms, commissions and advertising. Within just two months, their decision paid off. Financial reports from a January 29, 1935 meeting indicated a marked increase in lot sales, which in turn, enabled the board to rescind the $100 employee pay cuts of 1934 and promptly restore previous salaries. Members commended the men who, behaving as everyday heroes, sacrificed financially during a time of nationwide economic hardship to maintain and preserve historic Ferncliff.

PICTURE THIS

A cemetery photo booklet soon became a reality with an initial 2,000-copy printing of "Ferncliff Beautiful"—a well-conceived and highly successful community marketing piece that captured all the rural splendor of Springfield's trend-setting cemetery. Winwood & Company, meanwhile, received a renewed contract for the year, but board members passed a motion to terminate the company's sales contract as of November 15, 1935, with commissions of all lot sales prior to November 1 issued to the company.

J.B. Cartmell was elected association president for the next several years, receiving a tremendous vote of confidence after so ably leading members through Ferncliff's period of financial hardship.

As the cemetery association strove to maintain the highest possible quality for its grounds, buildings and monuments, so, too, must superintendents and grounds crews preserve Ferncliff's aesthetic beauty by working within strict parameters. The cemetery's bylaws were thus amended by

Monument to honor WWII, Korean, & Vietnam veterans.

trustees on May 9, 1940 to state: "All memorials, mausoleums, monuments, markers and other permanent memorial structures must be constructed from first-quality American marble or granite from recognized quarried or standard-cast bronze of approved composition, all of which are guaranteed by responsible producers to be equal in both quality and finish to accepted samples on display at the cemetery office.

"Marble markers (and) head of foot stones may be used only to match larger memorials placed on the same lot. Producers of monument materials, meaning, thereby, quarries, quarries (which) also manufacture memorials, and manufacturers of memorials, not quarries, must, in order to secure the approval of the cemetery, file with the cemetery an agreement to sell only first-grade, clear stone for memorial purposes, unless guaranteed that such stone is free from sap or anything which will cause attaint. That it will not check or crack, and that should any such defects develop within 10 years of the date of setting, the memorial will be replaced without cost to the cemetery or the lot owners by such quarrier, so manufacturing such memorial or by the manufacture thereof, who shall make satisfactory adjustments, such adjustments not to delay the replacement of the memorial and the cemetery.

"Any producer or retail dealer who violates this regulation shall be removed from the list of approved producers and retail dealers. The design, size and material for memorials must be approved by the officers of this cemetery, and all implications for foundations must be presented on a regular form provided by the cemetery and be signed by both the lot owner and the dealer furnishing the memorial."

Such a carefully worded legal statement guaranteed the utmost in memorial-stone quality was utilized by everyone involved. Each and every detail remained non-negotiable regarding this most sensitive subject.

CARTMELL STEPS ASIDE

During an annual meeting of January 28, 1941, president J.B. Cartmell, who had served the board continuously for seven consecutive years, expressed his desire that the office of association president be filled by a board member who had yet to serve in this capacity. Members respected his wishes and nominated S.A. Bowman, who was elected unanimously. Stanford J. Perrott continued as superintendent, assisted by Charles M. Wilkerson.

Ferncliff Cemetery Association conducted important, but seemingly routine business in 1942, but with an annual meeting of January 26, 1943, elected J.B. Cartmell as president instead of Bowman. Curiously, there appeared to be no obvious reason for this abrupt change in leadership. Perrott and Wilkerson continued in their respective capacities as superintendent and assistant superintendent. Recommended was that the association purchase a 29-acre tract of land owned by DT&I Railroad Company, located west of and adjoining the cemetery, for a price not to exceed $300 per acre. Cartmell was empowered by the board to proceed with the purchase. He announced on January 25, 1944 that final proceedings regarding the addition of DT&I property were completed, and also reported that a contract had been signed by Clark County commissioners to establish a World War II veterans burial section, with half of its purchase price paid April 16, 1945 and the balance by October 15 of that year. In a show of support for the war effort, the cemetery association invested $50,000 of cemetery funds in Series G, U.S. government bonds. Over the next several months, the association further invested by purchasing substantial increments of U.S. government bonds.

Aside from rather routine business in 1945 and 1946, a new association president, Earl W. Fulmer, was elected ... and cemetery workers again requested an increase in pay. Superintendent Perrott presented a petition on October 7, 1946, addressed to him and signed by cemetery workers. Likely mindful of past negative experiences in this area, board members carefully considered the proposal because of the way in which this sensitive subject has been presented by a tactful Perrott. He reported to the board that he first called all workmen together to discuss their initial request: 15 cents more per hour, they said, would be sufficient. After gentlemanly, yet firm negotiation, employees settled for 10 cents an hour, which was granted October 1 following an income and expenses study of the previous year.

A Great Loss

A Great Loss

FERNCLIFF
MOURNS EIGHT
NOTABLES

He who has gone, so we but cherish his memory, abides with us, more potent, nay, more present than the living man.

—Antoine de Saint-Exupery
(1900-1944)

The Association confronted several unexpected vacancies during a January 28, 1947 annual meeting, lamenting the death of Harry S. Kissell and the resignation of longtime trustee Robert Lupfer. Yet as much as these men were missed by the association, given their unwavering belief in—and commitment to—cemetery association work, the passing of former president Joseph B. Cartmell affected board members in a deeply personal way. A man of considerable business savvy, Cartmell had lifted Ferncliff from the burden of debt during lean economic times, and by keeping the cemetery operational, thereby preserved its historical, rural legacy for generations to come.

Trustees called a special meeting on April 21, 1947 to pass a lengthy resolution commemorating his life: "It is with profound sorrow that we are called upon to record the death of Joseph P. Cartmell, a member of the Springfield Cemetery Association from April 4, 1899 to the time of his death, which occurred in his home in Springfield on April 15, 1947.

"During his nearly half century of continued service, he served as a member of the board of trustees, as a member of the executive committee, and for many years, from time to time, as president of the board of trustees. Mr. Cartmell, for a number of years, has been considered Springfield's foremost citizen, and expressions of highest praise for the living man and deepest grief on account of his death have been universal.

"As a constructive force in financial matters, he had few equals. With keen perception, indomitable courage, and unbounded confidence in the future, he was a natural leader. His great ability and clear judgment had much to do with the financial strength and successful operation of Ferncliff and of many other interests with which he was connected. We, his associates of the Cemetery Association, mourn his departure. We loved him and have had many evidences of his affection for us. He frequently spoke of his connection with Ferncliff in terms of affection and pride. He was a regular in attendance at our meetings and was always seeking to promote the welfare and beauty of this scenic burial ground."

The resolution was signed by association president E.W. Fulmer. Once again, the passing of yet another great man was marked and preserved for the ages.

Still affected by the loss of two prominent Springfield citizens, trustees gathered on January 27, 1948 and recorded the following: "The report of superintendent S.J. Perrott, which had been prepared shortly before his death (and) which covered operations for 1947, was presented by Charles M. Wilkerson, assistant superintendent. The financial report was also placed in the hands of association members, with an analysis of the same by Mrs. (Mabel K.) Chatwood, assistant secretary-treasurer."

A third friend of the cemetery was gone, and Stanford Perrott's death left a considerable void among the ranks of his admiring coworkers, as they loved and respected him beyond reproach. To the end, his legacy remained one of supporting his colleagues while respecting and honoring the association—tactful in any and all Ferncliff-related endeavors.

Charles Wilkerson was named acting superintendent until further action could be taken. In the meantime, association members honored his civic contributions by drafting a written memorial: "Stanford J. Perrott came to Ferncliff on the fourth of April, 1916, and shortly thereafter was made superintendent of the cemetery. In 1941, he was made secretary of the board, and thereafter he served as secretary and superintendent until his death on January 22, 1948.

"Mr. Perrott had experience in cemetery care prior to coming to Springfield, having been connected with Woodlawn Cemetery in Toledo and Woodland Cemetery in Dayton, Ohio. He was recognized throughout the state as an efficient cemetery manager and served as president of the Ohio Association of Cemetery Superintendents for the year 1921, and in 1936, that being the 50th year of his existence, he served as president of the American Association of Cemetery Superintendents.

"Mr. Perrott was especially qualified for the position of superintendent of the cemetery. He liked people and had the faculty of getting along with them, and had a real sympathy for those in trouble. He appreciated the beautiful in nature and early on began to make plans for beautifying the cemetery. He introduced many varieties of blooming trees and shrubs as, for example, flowering crabapple, pink and white dogwood, Hawthorne, cherry and redbud, which, when in bloom, made the cemetery a wonderland of nature. The natural beauty of the lower entrance was enhanced by the planting of snowdrops, trilliums, jonquils, hyacinths, tulips, and irises and crocuses, and for several months during the early spring and summer they make this entrance the outstanding one in the country, if not the world.

"Gentle and kind, courteous and considerate, efficient and industrious, dependable and respected, Stanford J. Perrott was… the grand dame of gentlemen. That a record be made of the appreciation of the life and character of Stanford J. Perrott, it is ordered that it be spread upon the minutes of the board, and that copies thereof be sent to members of his family."

BRANCHING OUT

During the 1940s, the cemetery association began offering loans to Clark County-based individuals and companies interested in building homes or developing subdivisions. Interest rates of four to five percent were commonly awarded and payments were made quarterly. Association officers were authorized to make account withdrawals from various building and loan companies to care for the loans, and due to their generosity and business savvy, many within the community were able to finance at an affordable rate of interest on a wide range of projects. As had been its tendency, the association continued to operate within the best interests of Clark County residents.

By the end of the decade, Charles M. Wilkerson was appointed Ferncliff superintendent during an annual meeting of January 25, 1949, taking over for the late Stanford Perrott. Sadly, his tenure was short-lived. On June 3, board president E.W. Fulmer authorized an appointment committee to prepare a resolution of death for Wilkerson, whose May passing shocked and saddened the Ferncliff family who'd elected him superintendent just four months prior.

The committee decided to assume all burial expenses and to extend his salary to June 1. Once again, an untimely death signaled that a capable successor be found. The board considered four men: locals C.E. Gordon (23 years' experience at Ferncliff), S. Van Bird (one-and-a-half years' experience at Ferncliff), and Lester Noble (no experience), along with outside candidate Orrin Perrott of Indianapolis (12 years' experience in cemetery administration). Full consideration was given to each prospect, with the following decision rendered: "We recommend, by unanimous vote, the appointment of S. Van Bird of Springfield, at present employed in the management group of Ferncliff, to be acting superintendent of Ferncliff Cemetery." He became its official superintendent by a unanimous vote on June 7, 1949.

With compassion, the executive committee allowed Mrs. Wilkerson to continue living in the superintendent's residence until August 1. She responded with a note of gratitude: "This formal acknowledgment of the lovely flowers you sent Charles does not express adequately our appreciation of your kindness. His pleasant associations with the board over a period of many years was a source of great satisfaction to him and he treasured your friendship. You have been very considerate and kind to me during this difficult time.

On June 7, 1949 E.W. Fulmer and the Springfield Cemetery Association unanimously adopted and recorded an official statement of grief and loss, thereby preserving the extended contributions of Charles Wilkerson. It read: "The trustees of the Springfield Cemetery Association record with sorrow the death on the 18th day of May, 1949 of Mr. Charles Wilkerson, superintendent of Ferncliff Cemetery.

"Mr. Wilkerson served the association for 33 years, first as an assistant superintendent and engineer, then advancing to the position of superintendent in January (of) 1948, upon the death of former superintendent Stanford J. Perrott.

"Mr. Wilkerson, a civil engineer by trade, was a devoted member of the First Lutheran Church, a past master of the H.S. Kissell Lodge of Free and Accepted Masons, a member of Springfield Chapter #48 (of the) Royal Arch Masons, and Springfield Council #17 (of the) Royal and Selected Masters. He was a respected citizen with many activities.

"Mr. Wilkerson's ability, sound judgment, and his ever-present desire to promote the interests of Ferncliff Cemetery have been of great service to this association, and his never-failing kindness, his frank, cheerful nature and his capacity for friendship endeared him to his associates. With a real sense of personal loss, the cemetery association causes this expression of his appreciation of his qualities, and of its deep regret at his death, to be entered into the records,

The Deffenbach Monument, Section D.

and directs that a copy be sent to Mr. Wilkerson's family with assurance of deep sympathy in their loss."

The association's high regard for its members and recognition of their sacrifice and expertise have been continually revealed in a recorded history that dates from its earliest, hand-written minutes to present-day, computerized operations. The practice of resolution writing has marked the passing of those who've contributed to Ferncliff, whether through its beautification, expansion or economic survival.

Once more during the decade, a notable man had died.

During a January 23, 1950 meeting members acknowledged with sorrow the November 19, 1949 death of Allan E. McGregor: "He gave to this association 30 years of faithful and unselfish service, and no matter, was never more conscientious and diligent in the performance of this duties or more thoughtful of the best interests of beautiful Ferncliff. His good judgment and intelligence combined with his willingness to advise made him a strong and reliable trustee.

"A gentleman in the best sense of the word and of the highest integrity and moral courage, he was endeared to all those who were privileged to know him. As his associates on this board of trustees, we deeply appreciated his sincere, straightforward character, and each of us feels a deep sense of personal grief at the termination of his pleasant association of so many years. His loss is keenly felt by the association, his officers and trustees, and we extend to his family our deepest (sympathies) in their bereavement." The official condolence was signed by association president E.W. Fulmer.

S. Van Bird, acting superintendent since the May, 1949 death of Charles Wilkerson, was named full superintendent, and immediately during the same meeting suggested that the board of trustees hire a general foreman to assist him in his duties. They agreed, provided a suitable candidate could be found.

ONGOING TRANSITION

Progressive citizens with keen business sense continued throughout the 1950s to approach the board regarding loans for prospective, outside building projects. Local contractor Don Six approached the board to procure a loan for construction of residences in the Meadow Lane area, a then undeveloped subdivision east of town off old Route 70 (present-day Route 42) and one mile east of Community Hospital. Six applied for the loan, but the board deferred its decision until further consultation with him could be arranged. The prospective residential area consisted of thirty lots.

The executive committee met Dec. 29, 1950 and agreed to grant Six his requested loan. Once again, the board exhibited its generosity and trust in local people of vision with insights into improving the area. In turn, average community residents could now afford to build a life for their families in post-war America.

Interestingly, it was also during this time that Mabel Chatwood, secretary to the board, became the first woman to serve in such a capacity, becoming an integral part of the historic Ferncliff story. Likewise, foreman Napoleon B. Wagner joined a cemetery team in transition. He was hired to assist Superintendent Van Bird, while James D. Hatfield was named assistant superintendent.

The annual meeting of 1952 produced a new association president, John McKenzie, who managed the usual business of overseeing the sale and purchase of stock, approving loans to the public, and improving Ferncliff's grounds and facilities. McKenzie directed E.W. Fulmer to draft a resolution in memorial tribute to longtime association member Hugh R. McCulloch recognizing his thirty four years of service. The resolution read, in part, that McCulloch "was a staunch and understanding friend of the association and a dependable counselor to his officers and staff. He was of a friendly, unassuming disposition. He liked people, and people with whom he came into contact liked him. Hugh was a gentleman always, in the highest meaning of the word. As a citizen and successful merchant, Mr. McCulloch was conscientious and fair-minded, giving freely his time and money to many constructive and worthwhile community projects."

The resolution concluded by emphasizing McCulloch's unending devotion to the cemetery, recalling "the inspiration and support that he brought to the deliberations of the association and its various offices and committees upon which he so faithfully served." As with those serving before and since, McCulloch and his faithful civic service have been unseen forces behind the continued success of Ferncliff Cemetery Association. A grateful town and its people remain, to this day, indebted to their visionary contributions.

CHAPEL SERVICE

In perhaps one of the sadder chapters in Ferncliff's history, discussion in January of 1953 turned to a neglected and boarded chapel unfit for use. In the short span of four decades, the once celebrated chapel had fallen into disrepair, and trustees empowered Van Bird to contact a wrecking company and determine the feasibility of removing the aging structure in exchange for salvage material. Their concern also extended to needed renovation of the administration building's interior.

Mysteriously, an executive committee meeting was held on February 17, 1953—just one month later—to confront Van Bird's apparently unexpected letter of resignation, effective April 15. No explanation was given for the superintendent's departure, and Mabel Chatwood, employed by the association for seventeen years as a bookkeeper and secretary, was appointed acting superintendent by a unanimous vote. She was officially named to the position on January 18, 1953, marking another historic first, and earned a salary in line with those of her predecessors. Joseph Hatfield was named Chatwood's assistant, and he moved in to the Plum Street residence originally built for founding superintendent, John Dick.

THE MOURNING CONTINUES

Once again, a letter of condolence dominated an annual meeting, this time on January 27, 1954. The note of sadness commemorated the life and death of Clarence J. Foster, a two-time board member recognized as a benefactor to the entire community of Springfield. A noted industrialist, church leader and philanthropist, educated by local public schools, Wittenberg Academy, Wittenberg College and the Nelson business school, Foster became a giant of the road roller industry, manufacturing roughly 75 percent of the world's production.

"Many boys and girls in this community will forever be grateful to Mr. Foster for assisting them in obtaining an education," the record stated. "His philanthropies were personal, quietly given, with no thought of public acclaim being accorded him. Mr. Foster was a great lover and devotee of clean sport, always finding time to support local sporting events, not only by expenditure of his own funds, but by attendance at all worthwhile events. He always manifested enthusiasm for American Legion Junior Baseball, making liberal contributions thereto from time to time so that the local teams could have proper equipment and management.

"He was a zealous and devoted church man, his father being one of the founders of our present High Street Methodist Church. Mr. Foster served honorably and worked for many years as a member of his board of trustees and his finance committee. He attended church regularly, taking an active interest in all of its several activities, and was, at times, most liberal in its support. Always a gentleman, he was ever most courteous, gracious and kind to his friends and acquaintances, always ready to advise with young people concerning their problems. He was a prodigious reader, well-versed in local, state and national problems, always able to discuss intelligently political, economic and financial situations, and did not hesitate to expound his views thereon.

"After his retirement, his city being in dire financial condition, he accepted without compensation a position on our citizens' committee, which ... after extensive consideration of the problem, recommended to the city commission that it enact an income tax ordinance—which was concurred in the commission, and as a result, the city has been able to pay off all its old past-due obligations and is now operating well within its budget, having thereby been enable to purchase much-needed equipment (and) furnish proper municipal services to its citizens—all of which were made possible by reason of Mr. Foster's committee.

"From the time of his initial election as a trustee of the Springfield Cemetery Association in January of 1934, and his election as vice president and member of the executive committee in 1946, he has given his position with this board his serious and undivided attention..."

Colleagues hinted at Foster's "highly esteemed and honorable" standing within and around Springfield, referring to him as the association's "most distinguished and valuable member."

Describing his death as "a great loss" for the city and county, they concluded that Ohio "has lost a gentleman and (a) devoted citizen."

It was during this time that John P. McKenzie was reelected association president. Frank Dock (vice president), A.G. Samuelson (treasurer) and Mabel Chatwood (superintendent and secretary) were reelected as well.

By June 28, 1955, another of Ferncliff's faithful servants was gone. Vice president Dock called a special meeting to acknowledge the sudden death of President McKenzie, a Springfield High School and Wittenberg College graduate, just eight days prior. The board unanimously selected Dock to assume the office of president, with Irving Brain serving as vice president. McKenzie's memorial resolution read, in part, "... As a member of our board of trustees, and for the past three years thereof, he gave unstintingly of time and efforts, and was an excellent leader, proposing many commendable improvements to our board, which were adopted and placed in effect. He was a man of sterling character. Always a gentleman, he was ever most courteous, gracious and kind to his friends and acquaintances, and endowed with a sense of humor beyond that of the average man, and stricken as he was during his last illness, he maintained the above qualities until his death."

Described in rich detail as a devoted family man, he was a citizen deeply and obviously missed.

On January 24, 1956, trustees elected by acclamation A.G. Samuelson to succeed Frank Dock as president, with Irving Brain succeeding himself as vice president. Curiously, the business of the board intensified in late May, as Samuelson presented Mabel Chatwood's letter of resignation. She agreed to remain until a suitable replacement was found—no hint as to what prompted her departure.

The Bauer Monument, Section O.

Leveling the Field

RIGHTS GAINED, BUILDINGS LOST

These old buildings do not belong to us only—they belonged to our forefathers, and they will belong to our descendants unless we play them false. They are not in any sense our property to do as we like with them. We are only trustees for those that come after us.

—William Morris,
founder of The Society for
the Protection of Ancient
Buildings (on principles of
conservation and repair,
1889)

Ferncliff Cemetery

Comfort station, built in 1928

Veterans organizations approached the board on June 14, 1956 requesting permission to erect a regulation cross atop each of Ferncliff Cemetery's war mounds, provided they agreed to "keep them painted and in good condition." Louis Shelton, a Biltmore, North Carolina native, was hired as grounds superintendent and scheduled to begin work September 1.

More importantly, a national labor movement had spread to the increasingly industrial town of Springfield, and by early July, Vice President Irving Brain presented a copy of a letter he authored and sent to Samuelson outlining the possibility of cemetery association employees joining a union. Wilbur Culp of Local 880, retail, wholesale and department store union of the AFL-CIO, met with the board on September 26, with Edward M. Rosenhahn, internal union representative, and Oscar T. Martin, legal counsel representing the cemetery association, among those present. An agreement between the two parties and prepared by Martin was entered into the slate, to be in force and effective from September 24, 1956 to September 23, 1957 and self-renewing for yearly periods thereafter, unless notice was filed as set forth in the agreement.

A legal format for negotiating employee compensation now firmly in place, the board's attention by January 29, 1957 returned to rather routine and necessary cemetery maintenance. Brain recommended during the annual meeting that repairs be made promptly to sunken graves and receding monuments, despite their significant expense. An area north of the administration building was to be converted to a park-like setting featuring non-monumental, flush-type markers, which could be sold at a higher price than the other locations. Such area improvements well-served Springfield's citizens as the board continued to upgrade every aspect of already-notable Ferncliff.

With worker contract expiration looming, a special meeting was held on September 18, 1957. With Local 880 representatives present, Ferncliff employees asked that all Saturday work be eliminated, and that they not be required to work outside during rain or other inclement weather except in an emergency. They further requested that the association place into effect an employee pension program covered by the bargaining unit, and that wages rise from $1.60 to $1.80 per hour. The cemetery association countered that elimination of Saturday work wasn't feasible, given the reality and unpredictability of death. Interments must be completed in the event of bad weather, but employees would now receive two hours' reporting pay. Although board members reported that a pension plan was not economically feasible, they offered a pay rate of $1.70 per hour, effective September 24, 1957, for a period of two years.

A new contract later drawn by Martin and presented on October 15, 1957 raised hourly wages to $1.70 for the first year and $1.72 for the second year, an approximate 5 percent increase.

LAND DEALS

During an executive meeting of the cemetery association on April 7, 1958, superintendent Shelton presented a plat suitable for 128 burials and prepared in accordance with a request by the Knights of Pythians Lodge, which remained interested in purchasing additional ground for burials. The ground would be held until December, pending a purchase decision. By November, an offering price of $91.20 was offered, which included a two-piece vault, conditioned upon the lodge purchasing a two-year option on a triangle-shaped piece of ground adjacent to the lodge's 128-space lot. Future cemetery planning necessitated that Shelton contact Ray Wyrick for cost estimates regarding road layout and drainage on approximately 100 acres of undeveloped land.

Shelter house

The association further agreed on August 13, 1959 that should Springfield desire to build a Plum Street fire station, members would sell a parcel of Plum Street land for this purpose at the property's appraised price, providing the city razed the first Plum Street house north and adjoining the cemetery and incorporated that site into any planned fire station property.

By October, superintendent Shelton reported that the following conditions were agreed upon and/or requested by Ferncliff workers: 1) an increase in the employee hospital plan with inclusion of dependents, 2) a retirement age of sixty-five with acceptance of board approval, 3) loss of a full vacation if an employee failed to work for a full year, and 4) no increase in salary for the year.

Of note, David L. Babson & Company was, at this time, the investment company accepted by the cemetery association to manage investments. David Porter of the David L. Babson Company served as financial consultant. Periodically, brochures were printed and circulated outlining various investment portfolios.

THE 1960s

In February 1961 Superintendent Shelton presented a plan for rebuilding and improving Ferncliff's maintenance buildings, but upon board recommendation, he was asked to prepare plans for new buildings in time for the next board of trustees meeting, as this approach was considered a more satisfactory and lasting solution. Trustees approved in March that final plans would be drawn by a local architect and submitted for bids.

The 1960s will long be remembered, among other things, as the era when America was gripped in the spirit of "urban renewal." Old buildings and structures were cast aside as a fever for the new and modern, a trend which had started in the 1950s, continued to sweep the country. Ferncliff was not immune. With construction of the new also came demolition of the old, and by April of 1961, Snyder Park trustees expressed interest in purchasing the stones from Ferncliff's original chapel and were willing to demolish the structure in order to obtain them. *(See photo on page 60)* The stones from the once majestic chapel were used to construct park shelter houses. Likewise, plans for a new maintenance building to replace Ferncliff's vault and equipment storage were finalized.

In July, plans were also made for the demolition of a restroom and shelter house, which was once utilized by citizens who awaited street cars. Understandably, a debate raged regarding the "whys and wherefores" of demolishing historic structures, but in the end, the decision was to tear down instead of preserve.

On May 8, 1962 money was appropriated to remove Ferncliff chapel's old foundation to approximately three feet below ground level, in order that the entire area be filled in, sewn with grass, and appear as it does today. Four months later, Shelton brought to the board's attention that the following year marked Ferncliff Cemetery's 100th anniversary and suggested planning get under way to publicly recognize the centennial during the upcoming Memorial Day. Irving Brain was elected association president in late January of 1963 and entrusted with the added responsibilities of overseeing Ferncliff's milestone celebration. The anniversary was commemorated at the conclusion of the annual Memorial Day parade with Congressman Robert Taft, Jr. offering

Ferncliff Mausoleums.

remarks from the GAR mound in Ferncliff. The occasion, which was attended by 5000 people, was also marked with a Color Guard, patriotic music, prayers, and a recitation of the Gettysburg Address by a Springfield High School student.

The year 1964 began with the annual meeting of the Springfield Cemetery Association held on January 28, with the discussion led by Paul Deer, the founder of the Bonded Oil Company, concerning the perpetual care fund which allows for the oversight of the gravesites. His motion, which carried during the executive session, indicated that the percentage set aside for the fund should be increased to 35%, which would mean that 35% of the money collected from lot sales would be placed into that particular fund, an increase of 15%. An amendment to the constitution needed to be made at the next annual meeting in order to legally finalize the official business of changing the document.

During the executive meeting called for February 11, the board listened as Superintendent Shelton presented a drawing of the proposal for a Veteran's Memorial that would be placed in the cemetery when the veteran's group raised the necessary funds. Before the next meeting was called, board president Irving H. Brain passed away leaving a major void in the proceedings of the association. A memorial to Mr. Brain acknowledged his twenty-three years of service to the association giving his "good judgment and intelligence which made him a strong and valuable member of the Association."

Newly elected president John D. Kuhns conducted the executive board meeting held July 14. The discussion of the Veteran's Memorial continued with the approval to extend permission to the Veteran's Administration that appropriate lighting could be tastefully placed illuminating the flag pole at their own expense with the cemetery paying the electrical expenses. The monument-style rostrum placed in the front yard before the administration building serves as the main speaking platform for invited guests as they commemorate and conclude the special Memorial Day events held each year.

President Kuhns presided over the annual meeting January 26, 1965, with the usual business conducted and concluded, but with a sad note informing the association members that assistant superintendent Joseph D. Hatfield had passed away. Harold Pendleton was approved and appointed to replace Mr. Hatfield. The trustees decided to provide for Mrs. Hatfield a pro-rated salary, two grave spaces, and the use of the residence on the cemetery grounds until notified otherwise by the trustees giving the family at least a sixty day notice.

The 1966 annual meeting called by president Kuhns finalized the motion to the change in the constitution set before the body during the meeting in 1964 concerning the increase in the percentage of lot sales to be set aside for the perpetual care fund. The association quickly approved the 35% increase to be retroactive from the year 1965. Other than a vote to improve and extend the union contract for three years, improvements to the roadways, designing several proposed new sections, and routine business, the everyday workings of the cemetery continued uneventfully. The same could be said for the next several years.

At the annual meeting of January 19, 1971, Association President Kuhns alerted members to the fact that the Ohio State Highway Department of Transportation intended to acquire cemetery property adjacent to West First Street in order to widen State Route 41. Several years of negotiations ensued over a fair price for the land, but on November 12, 1974 the superintendent informed the Executive Committee that the State of Ohio had agreed to meet their asking price of $130,000.

As time passed the members of the Board of Trustees recognized the importance of "keeping up with the times" in that, in order to remain competitive with other cemeteries, Ferncliff should offer above ground burial. During the meeting of August 9, 1977, Superintendent Shelton was asked to obtain estimates for a ninety-six crypt mausoleum, something that had been discussed on several occasions. Eventually the board would approve plans to construct a 124-crypt mausoleum and to convert the garage area of the office building to a stand-up chapel. The mausoleum was finished on November 14, 1978 with the first interment on January 24, 1979. The business of the Association continued to progress as usual with the routine of meetings and necessary maintenance issues with very little out of the ordinary concerns involving the everyday function of the cemetery. One item of note observed in the minutes of the board meeting of February 11, 1986, concerning the World War II veteran's mound was an issue introduced by Superintendent Shelton that the venerable ground set aside for the burial of that war's veterans would be filled in three to four years. The superintendent was to meet with the Clark County Veteran's Service officer to discuss the various possibilities for another veteran section.

The next entry into the agenda dealing with the annex to the World War II veteran's mound reported on May 12, 1987, that the board moved, seconded, and passed a motion to sell to the Clark County Commission twenty-five spaces for the new war mound to be located across the road from the present area when completed. The Clark County Commission met with the board on July 14, 1987, with the proposal to buy an additional fifty grave spaces on the annex. The board unanimously approved the proposal to increase the size of the annex in anticipation of more space being necessary in the future. Grading of the annexed area to the mound began in the fall of that year.

Superintendent Louis Shelton announced his desire to retire with the reading of his letter of resignation on September 11, 1990, leaving quite a void as he had led Ferncliff Cemetery through some very productive years. Effective October 1, 1990, the executive committee named Harold Pendleton as interim superintendent.

Eventually, the board removed the "interim" title from Harold Pendleton's job description as he began his smooth transition as Ferncliff Cemetery's newly selected superintendent. The naming of his assistant was tabled until a later time. That time came quickly as, during the December 10, 1991, executive committee meeting, the board named Stanley Spitler as assistant superintendent, a decision that would eventually prove to be very wise.

Ferncliff Cemetery continued to expand its space for above ground burials. The board of trustees introduced the idea to construct another mausoleum building at the June 14, 1994, meeting. During the August 9th meeting, Superintendent Pendleton presented plans for a new 80 crypt mausoleum building to be constructed of the same materials as used for the mausoleum built in 1978. As the 1990s wore on, the cemetery association conducted the usual amount of business in the improvements and overall care of the cemetery grounds. As time would progress, several venerable members of the association passed away, and nominations of new members were presented to take their place, members who would lead the association into the 21st century.

בית מועד לכל חי

קאנגרעגיישאן

חסד של אמת

תר'פ'ב

HEBREW CEMETERY

One of two Hebrew sections in Ferncliff.

A New Era

AN ELEGANT VISION

Cemeteries and graveyards face dramatic pressures—from development, from abandonment and decay, from nature and from vandals. In many respects they are also very different from other historic and archaeological resources, since they often involve a variety of functions—sacred, artistic, historical and genealogical. The resources present include not only the human remains, but also the sculptures and monuments, as well as the landscape itself, making cemeteries—and their preservation —very complex.

—Chicora.org,
Chicora Foundation, Inc.

Early in the new century Ferncliff Superintendent Harold Pendleton announced to the board of trustees his desire to retire on December 31, 2002. Pendleton had served the Springfield Cemetery Association for 43 years, the last twelve in the position of superintendent. A reception was held at the Springfield Museum of Art on December 17, 2002 to honor the outgoing superintendent and pay final tribute for his dedicated and longstanding service to Ferncliff Cemetery.

A search committee was appointed to select a replacement for the retiring Pendleton. On November 12, 2002, the committee reported that the current assistant superintendent, Stanley Spitler, was their recommendation to the trustees. At that time, the trustees voted unanimously to name Spitler as the new superintendent of Ferncliff Cemetery following Harold Pendleton's retirement.

Ferncliff would thrive under Spitler's leadership. With the approval of a farsighted board of trustees, he began a program of improvements that would usher graceful, elegant Ferncliff into the twenty-first century. His first order of business was to oversee the complete renovation of an administration building that dated to 1930. Next, he gave priority to computerizing burial records and plot-finder systems in an effort to assist historians, genealogists, descendants and other interested persons with interment location and information. The board of trustees' and the superintendent's future plans for Ferncliff include the addition of walking trails, a cremation-scattering garden, a community facility, additional mausoleums, family estates, and perhaps a new chapel.

The ongoing commitment to beautification of the cemetery by the Springfield Cemetery Association, the board of trustees and Spitler was rewarded when Ferncliff Cemetery was renamed Ferncliff Cemetery and Arboretum in 2006. As part of the process to be named an arboretum, a tree inventory was undertaken in 2004 and 2005. Trees were tagged with both their botanical names and common names. Approximately twenty-five to thirty trees are planted each year to replace those that have died, or to enhance areas not yet developed. Ferncliff also gained the distinct privilege of having a nationally recognized tree propagated for community planting and enjoyment. Officially named the Diamond Bark American Beech Ferncliff, the crossbred species boasts diamond-shaped, rough bark, unique when compared to the more typical smooth bark of American Beech trees.

IN THE LUDLOW TRADITION

Originally, cemeteries were designed not just as burial sites, but as parks —places where visitors arrived in their horse-drawn carriages with picnic lunches to enjoy while visiting deceased loved ones. In the era before public parks, cemeteries designed in the rural tradition provided a place to retreat from the stress of everyday life, a place for quiet reflection. Cemeteries were also places of beauty where art and nature married. Visitors enjoyed not only the ethereal beauty of funerary art, which was

Ferncliff Cemetery

wrought by man, but also the quintessential beauty of each season, courtesy of nature. Ferncliff pioneers such as John Ludlow envisioned their cemetery as a lush, green, beautiful gift to the Springfield community. As more than a century has passed since Ferncliff's founding, it has also become an unparalleled repository of Springfield history. Much more than a graveyard, a cemetery represents a chain by which families link and extend themselves and their generations. These "gardens of stone" are, in fact, physical manifestations of civilization's most dignified rituals -- the honoring of lives whose worth shall not be forgotten. Each tombstone documents the social, cultural, military or political history of a community. Monuments that rise from hallowed ground are at once repositories for the past and inspirations for the future. Because of them, men know from whence they came, and can move forward, keeping in mind those who made the way possible.

Man has an innate desire to be remembered for the life he led while present upon this earth. Ferncliff Cemetery provides that place of permanent memorialization. Springfield's magnificent, rural cemetery has dedicated itself, through instinctive and calculated guidance, to protecting and preserving the lives and memories of its people. Each individual who rests there in peace remains important enough to be forever remembered for his or her unique sacrifices and contributions to family, friends and community.

Each generation of Ferncliff's leaders have also reaffirmed the commitment to making it a wonderful place in the city for people to visit. Long range plans for upgrading and improving cemetery grounds, and a commitment to sharing the abundant beauty and rich history of Ferncliff with the public, will continue to make Ferncliff a gift for the community to enjoy for generations.

Stone Stories

CEMETERY SYMBOLISM UNVEILED

"A good character is the best tombstone. Those who loved you and were helped by you will remember you when forget-me-nots have withered. Carve your name on hearts, not on marble."

—Charles H. Spurgeon (1834-1892), English Baptist preacher, author and editor

Visual symbols and eloquent phrases remain cemeteries' grandest of story-tellers. Perhaps nowhere is symbolism richer or more apparent. Preoccupation with death during the 17th and 18th centuries has morphed into a softer, gentler form of contemporary mourning and remembrance that is becoming increasingly individualized. Traditional weeping willows and winged cherubs are giving way to life-like laser portraits and skilled renderings of occupational, recreational or philanthropic interests. Today, a monument's decor is limited only by one's imagination and budget.

According to Doug Keister's *Stories in Stone*, "Grave markers themselves are symbols, for they meet the most basic definition of a symbol: something that stands for and represents something else." Because tombstones represent the lives and personalities of individuals who once walked this Earth, their inscriptions and symbols provide insight not only into who they were, but the period in which they lived.

"It is possible to read in broad terms the greatest shifts in cultural values that have occurred over the course of time," Keister writes, "by closely examining the changing nature of those carved, sculpted, and engraved images." Interestingly, a majority of 17th- and 18th-century gravestones are adorned with stark symbols and emblems that focus on death and mortality—reflections, no doubt, of a societal preoccupation with the inevitability of death that existed during particularly harsh times. Cemeteries of that era, thus, are ripe with "symbolic representation of ... predominant cultural messages meant to continuously remind us of our impending end": skulls, skeletons, and winged hourglasses, shattered urns, sickles and coffins.

But "attitudes softened over the next several centuries," notes Keister, a noted photographer who's authored more than thirty books and specializes in historical cemetery and residential architecture. "Harsh mortality symbolism began to give way to a gentler form of mourning imagery." Skulls and skeletons were slowly replaced with praying hands, heavenly gates, crosses, flowers and lambs, marking a noted shift in society's perceptions of death. Instead of reminding graveside visitors of man's impending end, tombstone symbolism was, instead, beginning to suggest belief in an afterlife.

By contrast, modern cemeteries remain almost entirely void of such imagery. "It has most definitely taken a back seat to the astounding number and variety of highly personalized, retrospective symbols referencing occupational and recreational pursuits that are, in many ways, the most distinguishing features of contemporary grave markers," Keister says.

A look, now, at some of history's most popular gravestone etchings and symbols:

Anchor: The symbol of hope (Hebrew 6:19-20).

Balls: Mostly decorative, but in groups of three can represent gifts of money.

Book (Opened): A representation of the human heart. Thoughts and feelings; open to God and the world.

Celtic Cross: Often embellished with detailed symbolism associated with a person's heritage. Basic form: cross enclosed in a circle. Predates Christianity.

CHI-RHO: Oldest Christian symbol. Chi and Rho are the first two letters of the Greek word for Christ. Also used to symbolize a "good omen" or a "good thing" from original Greek.

Corn: Found in America's heartland and one of the oldest harvested crops. Used as a symbol of fertility and rebirth.

Cross with Crown: Christian symbol of the sovereignty of God. "Crown" as in "victory." The cross as representative of Christianity.

Dove: The most frequently viewed animal art in cemeteries. Many poses. Often seen holding an olive branch (Genesis 8:10-11). A symbol of purity and peace and the Holy Spirit (John 1:31-33). God made peace with man.

Eagle: One of the most powerful bird symbols. Also a symbol of America, resurrection or rebirth. It was once thought that an eagle flew toward the sun, burned its feathers and plunged into the sea, but was rejuvenated. (Psalms 103:4-5). An eagle's watery renewal is compared to baptism. The two-headed eagle is associated with power and respect. Two heads amplifies the point. One of humankind's oldest emblematic symbols.

Easter Lily: Purity, chastity. Casting off earthly things. Attaining heavenly, spiritual qualities. The plant boasts plain foliage but a beautiful flower. In Greek mythology, white lilies were droplets formed by the milk of Juno, the mother of all gods, after she created the Milky Way.

Fern: Humility, frankness, sincerity.

GAR: Grand Army of the Republic. Membership is limited to veterans of Union army, navy and marine corps who served between April 9, 1861 and April 9, 1865. Membership provided companionship through organized encampments. Established soldiers' homes and lobbied for pensions. Five U.S. presidents were GAR members. Instrumental in establishing Memorial Day, May 30, as a national holiday of remembrance. GAR no longer exists. A modern-day version is the VFW.

Gate: Represents passage from one realm to the next. Christ breaking through the barrier between the lost world and Heaven.

Grape Clusters: Symbols of Christ's blood and of changing Christ's blood into wine. Grapes combined with shocks of wheat symbolize Holy Communion.

Greek Cross: St. Andrews cross signifies suffering and humility. Maltese cross: forms eight points, representing the Beatitudes. Associated with heraldry worn by Christian warriors as they marched to the crusades. Floriated cross: arms end in three, petal-like projections. Broad-footed cross: arms look like triangles with the tips of four triangles pointed toward the center. These symbolize the protective power of the cross.

Hand (pointing upward): A soul risen to heaven; represents the hand of God.

Hands (clasped): Shaking hands are a symbol of matrimony. The presence of a sleeve depicts feminine and masculine. Gender neutral represents a heavenly welcome or earthly farewell.

IHS or **IHC**: Letters overlaid, often looking like a dollar sign. IHS derives from the first three letters of Jesus' name, using the Greek alphabet. IHC represents the first three letters of Jesus' name using the Roman alphabet.

Ivy: Its eternal greenness is often associated with immortality and fidelity. Its leaves cling to a support, a symbol of attachment, friendship or undying affection. Its three pointed leaves provide a symbol of the Trinity of God.

Lamb: Usually marks the graves of children. Symbolizes innocence. Also the most frequently used symbol of Christ. The Lamb of God (John 1:29).

Latin Cross: The cross most commonly associated with Christianity.

Laurel Wreath: In funerary art the laurel wreath typically symbolizes immortality or eternity. The association with eternity is due to the fact that laurel leaves do not wilt or fade. The wreaths are also a symbol of victory as winning athletes in the ancient contests were often crowned with them, which was supposed to convey immortality.

Menorah: A seven-branched candelabra seen on the tombstones of righteous women. The menorah is depicted in Exodus 25:31-40.

Oak Tree or **Leaves**: An oak is considered the king of trees, symbolizing strength, endurance, eternity, honor, liberty, hospitality, faith or virtue. All combined provide a symbol of the power of Christian faith during times of adversity.

Obelisk: A popular form of Egyptian architecture representing a ray of sunlight. First seen in Egypt in 2650-2134 BC.

Pebbles: It's a Jewish custom to leave pebbles on or around a tombstone. Coins are also used, as are bits of glass as substitutes, signifying someone still cares and remembers. A thousands-year-old tradition. The Old Testament has many references of stones being used to cover or mark graves. (Joshua 4:1-9). Stones are powerful symbols of the "people of Israel." Rock mounds were used as burial markers because the dead could not be buried very deep. Also helped protect a burial from animals. Jews were nomadic people. As they passed by a gravesite, they performed maintenance by adding stones to it.

Pineapple: A symbol of hospitality. Once presented as gifts to seafarers after they returned home from sea.

Rose: Queen of flowers due to fragrance, longevity and beauty. Inspires lovers, dreamers and poets. Christian symbolism: red for martyrdom, white for purity. Christian mythology: the original rose in Paradise had no thorns, but acquired them on earth to remind man of his fall from grace. Its fragrance remained, providing a hint of Paradise. Roses are frequently placed upon women's graves in Victorian-era cemeteries.

Skull: A symbol of death. No escape.

Star of David: Divine protection. The most well-known Jewish symbol. Didn't become a major symbol until the 1880s. Became a permanent identifier of Jews when Hitler ordered it worn on their armbands. Israel placed this symbol on its flag.

Stone: Refers to resurrection. Large stones are associated with great power.

Thistle: Curse associated with earthly sorrow; connected with the crown of thorns and the passion of Christ. Commonly used on gravestones of those with Scottish ancestry.

Torch (inverted): With flames burning, symbolizes death. Also suggests that the soul (or fire) continues to exist in the next realm. Upside down without flames represents an extinguished life.

Tree Stones: Derived from Victorian times, particularly 1880-1905. If one is found in a cemetery, often many more can be found. Symbolizes popularity and ties to nature. Could have likely been ordered from Sears & Roebuck, quite popular in the Midwest. People read the catalog and became familiar with style. Also popular with members of the Woodmen of the World fraternal organization.

Urn (Draped): The most common 19th-century funerary symbol. The drape represents reverential accessory or a symbol of the veil between earth and heaven. Urns symbolize ashes or created remains, but urns seldom contained ashes during the 19th century. Also used as decorative devices.

MOMENTO MORI

Although both the location and appearance of a cemetery can reveal clues about the individuals buried there, the most prolific source of information remains its gravestones. "Even at a distance a grave marker—its size, its complexity, the material it is carved from—can reveal certain things about the person buried under it, such as roughly when they died and whether they were wealthy or not," Keister says. But in addition to providing clues of pedigree, tombstones often provide much more detailed and personalized information. "Through inscriptions, epitaphs, and symbolic images, markers and memorials communicate the essential details of the deceased's life, as well as their thoughts and feelings, or those of the people closest to them."

New England Puritans, whose lives were often dangerous, diseased and difficult, "viewed death dispassionately as an unpleasant but unavoidable reality," Keister emphasizes. A thoroughly religious community, Puritans viewed death as the moment in which their earthly life is forever judged and the fate of their souls decided. Puritan gravestones, thus, reinforce the idea of individual mortality. Says Keister, "A skull and crossed bones, a winged death, and an hourglass—representing the swift passage of time—were popular visual motifs." Accompanying epitaphs, he adds, "state the matter plainly: 'As I am now, so you shall be.' " Their message was of Memento Mori, roughly translated from Latin and meaning, "Remember you are mortal" or "Remember you will die."

THE LANGUAGE OF SYMBOLS

Symbols possess what can be considered a universal language, bridging cultures, religions and eras to, as Keister writes, "tell the same story to different people regardless of their native language or level of literacy." Because of symbols' universal appeal, they have been relied upon throughout cemetery history to leave lasting messages easily understood by anyone who views them.

WHAT'S IN THAT STONE

Since the late 19th century grave markers have been consistently constructed of granite, a hard and attractive, igneous rock that resists weathering and wear. Prior to this time, however, tombstones were carved from the strongest materials available, making it relatively easy to date a grave by identifying the material used.

Wood or stones were a popular, early American choice—often inscribed with the name or initials of the deceased. Sedimentary rock such as slate or limestone followed and was utilized until the middle of the 1850s. Relatively soft, the rock wore easily and proved unable to withstand the test of time. A move was eventually made to stronger materials, such as marble or granite.

FERNCLIFF MAUSOLEUMS & STRUCTURES

The Bushnell mausoleum remains "a very well-executed, historically accurate Greek Doric mode of architecture," according to George Berkhofer, former Clark County Historical Society curator and executive director. The building features "a peripteral colonnade, with four columns across the front and six down the sides. Its classic attention to Greek Doric detail includes three steps located at the front. Classic Greek temples, in contrast, typically utilize steps on all four sides. Greek Doric architecture is perhaps most well-known and identified in the famous Parthenon.

The Gladfelter mausoleum is of typical late 19th-century design, with its center doorway featuring an Egyptian touch with a relatively simple lintel and two side pieces. Berkhofer identifies its surrounding stonework as "rusticated" because its surface has been left "in a rather rough fashion," and attributes its use as a "statement of strength and stability." (See pgs. 58–59)

Pure Egyptian conception, meanwhile, is a hallmark of the McGilvray mausoleum, with its "system of massive stonework consisting of large, squared blocks and panels. Its design features two front columns and two "normal, in antis columns" on either side of its opening—"simplified lotus columns." Its walls are constructed at a slant, in Egyptian pylon style. Above the engraved name "McGilvray" is a wide, decorative band bearing the symbol of the Greek god, Horace. Berkhofer describes the tomb as a simplified version of a "very standardized type of Egyptianesque architecture"—another sign of strength and stability.

Interestingly, the Blee mausoleum incorporates several characteristics of other Ferncliff structures. Berkhofer notes the "Egyptianesque effect" created by its sloping walls. "The sides of the building are like the Gladfelter, composed of huge blocks or slabs, virtually cyclopean in size, (and) of rusticated masonry," he says. Its doorway features a lintel with 'Blee' in raised letters, with "Greco-Roman triumphal garlands on either side." The side walls are finished in a highly stylized type of Greek pilaster that can also be described as a flattened-column motif. "There are four flutings on either side." Above them runs a "stylized flowering or basket-like motif, which (is) the equivalent of an actual Corinthian column capital."

Of a more medieval, Christian style is the P.P. Mast mausoleum, its walls constructed of rusticated stone and its columns fabricated from marble. On either side are curving, decorative lower walls that "terminate in stonework pedestals, with larger stone planters on top of them." The column capitals boast a medieval

appearance, with their leafy design "an obvious holdover from the art nouveau period (1890-1914)."

The Baldwin family mausoleum is built into the side of a hill—a construction technique Berkhofer identifies as commonly used during the time to control expenses. "At the face of the vault is a primitive Greek doorway with two uprights for the side. The two columns were originally fluted, but most ... has eroded over time." Behind the columns is a flat wall with an arched opening—a simulated keystone at its top. "The Greeks rarely ever used keystones, but the Romans loved them," Berkhofer says, identifying the architecture as Greco-Roman. The structure's entryway is surrounded by rusticated stone—a feature typically referred to as "dromos" by the Greeks.

The stately John Bookwalter monument is of Gothic Revival architecture, large enough to serve as a chapel or small church. Its walls feature rusticated stone and buttressing, most likely constructed from sandstone or limestone. Its front entryway forms a lancet arch, with the lower portion boasting two granite columns of elaborate Corinthian style. Halfway up the front pediment a trefoil window is carved into the stonework. Its roof is built of solid stone panels whose crevices are covered with stone moldings. The front peak contains a grand finial that springs from a carved piece of stone.

Ferncliff's administration building displays late Gothic Revival design in its stone buttresses and rusticated stone walls. Some of its windows "terminate in pointed arches," others are lancet, rectangular or square and reflect the Tudor style of the 16th century. The building, "is in tune with its period of construction, the 1930s," Berkhofer says, and has "eliminated a great deal of external decoration as a bow or nod to the art deco of the period, which tends to be heavier, massive and virtually Egyptianesque at times. It's a reflection of the increasing role of government and administration in all aspects of human life at that time." Because United States cemeteries traditionally bear at least some religious overtones, the most common being gothic, that style remains the overriding motif of the building. (See pgs. 74–75)

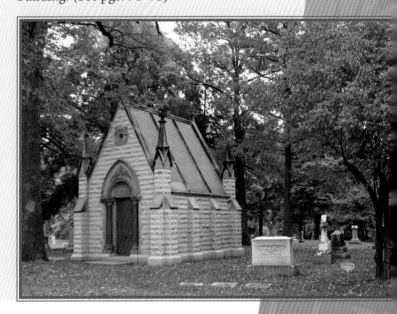

Roads Less Traveled

THE QUIET BEAUTY OF FERNCLIFF CEMETERY AND ARBORETUM

written by
Pam Corle-Bennett

He that planteth a tree is a servant of God; he provideth a kindness for many generations, and faces that he hath not seen shall bless him.

—Henry Van Dyke
(1852-1933)

Ferncliff Cemetery & Arboretum remains a special place for people who appreciate trees or enjoy a quiet, casual stroll to soak in nature's diversity, its sights and its smells. As Minnie Aumonier once wrote, "There is always music amongst the trees in the garden, but our hearts must be very quiet to hear it." Throughout the years, many dedicated superintendents have focused on diversifying the cemetery's growing species. Through diversification, a continual canopy covers Ferncliff's grounds. Nursing a diverse population ensures the entire cemetery won't suffer if problems occur to any one of its tree species. Generously, Ferncliff officials encourage the public to walk its grounds and enjoy its educational role as a meticulously maintained arboretum. Each visit is sure to enhance one's understanding of trees and their impact upon the environment.

Superintendent Stan Spitler gathered the visions of his predecessors and maintained their diligence in preserving Ferncliff's urban forest, developing an aggressive campaign of continual tree replacement. Those stately growths that have passed their prime or been damaged by storms or pests are quickly and efficiently replaced with new cultivars. Trees of significance are selected based upon their value or beauty to the cemetery, or because they remain underused in the area and are found within Ferncliff.

Of particular, stately importance in Ferncliff are Ohio Champion and Diamondbark trees. In 2005, while Superintendent Spitler was touring the cemetery's grounds with an area forester, the pair discovered an American beech with an unusual bark. The normally smooth, gray bark featured shallow fissures or depression. As the tree aged, its fissures deepened, resulting in cracks that formed a conspicuous, diamond-shaped pattern. Cuttings were taken from the unique tree and plans are to propagate it for future availability in the nursery trade. The cultivar was appropriately named 'Diamondbark.'

Champion trees are distinguished on the basis of their grand measurements, and Ferncliff boasts six plants deemed Ohio Champion trees, including the Crataegus mollis (Downy Hawthorn), Hemiptelea davidii (Hemiptelea), Ulmus rubra (Slippery Elm), Viburnum prunifolium (Blackhaw Viburnum), and Viburnum rufidulum (Rusty Viburnum), the latter two being of the Honeysuckle family.

Since 1940, the largest known specimens of native and naturalized trees in the United States have been documented. In Ohio, each nominee reported to the state's Department of Natural Resources receives a score based upon trunk circumference, crown spread, and total height. Ohio foresters and other experts evaluate the competitors and assign the 'Big Tree' honor, making Ferncliff's stable of Ohio Champions a notable achievement.

Other trees of prominence within the cemetery's fenced grounds have helped Ferncliff achieve its unique status as an arboretum, defined by *Merriam-Webster's Online Dictionary* as "a place where trees, shrubs, and herbaceous plants are cultivated for scientific and educational purposes," or, more simply, a botanical center devoted to the exhibition and study of trees. Ferncliff's most prolific include:

White or Concolor Fir (Abies concolor)

Most firs tend to perform poorly in Ohio clay or soils with problematic drainage and are usually native to mountainous habitats and well-drained, acidic soil. Although the concolor fir is native to mountain areas in Oregon, California, Colorado and Northern Mexico, it has endured Ohio conditions better than others. Its leaves are approximately two inches long and sport a pair of bluish-white bands on their undersides—a feature used to identify many firs. Crushing its leaves produces a slight citrus fragrance. Its foliage wears a softer appearance than spruces, and healthy specimens can grow 30 to 50 feet tall.

Hedge Maple (Acer campestre)

This medium-sized maple is not used enough throughout Ohio landscapes and should be used more often. The hedge maple grows 35 to 50 feet tall and boasts a wide, rounding spreading habit. At times, its width becomes comparable to its height. Summer foliage is relatively disease- and insect-free, and its fall color remains fairly nondescript—a ruddy yellow at best. It remains an excellent specimen tree for small landscapes and does well in a variety of conditions. The tree is commonly used in England for hedge rows, as it tolerates extreme pruning.

Paperbark Maple (Acer griseum)

A defining characteristic that makes this tree a favorite maple is its cinnamon-colored peeling and exfoliating bark. Its young stems turn from a rich brown to a reddish brown as it ages and its outer bark begins to peel off. Its leaves cluster in threes (trifoliate) and turn a beautiful burgundy, wine red in the fall. The paperbark maple grows slowly (to roughly 30 feet tall) and its shape is somewhat upright and oval or rounded. Very few insects or diseases are typically seen on this Ferncliff specimen.

Sugar Maple, Hard Maple, Rock Maple (Acer saccharum)

Native to Ohio, the sugar maple is the cemetery's dominant tree species and was most likely one of the first to fill the landscape during Ferncliff's early stages of development. It grows 50-70 feet tall and boasts exceptional late-year color (a blend of oranges, reds and yellows), making Ferncliff an excellent Clark County fall viewing location. Newer cultivars are being developed with such features as scorch-resistance, fall color consistency, and upright growth habits. Sugar maples are prevalent throughout Clark County's natural areas and widely distributed throughout Ohio. They're frequently used for tapping sugars that are later boiled to produce maple syrup.

Ohio Buckeye (Aesculus glabra)

This native tree is found in a few places inside the cemetery. Wherever the Ohio buckeye is located, passersby typically spot "buckeye nut" collectors beneath its fall branches. Hunters must arrive pretty early, however, to beat the squirrels. Buckeye fruits are covered by a spiny, light brown husk and despite being poisonous, remain their food of choice. Ohio buckeyes typically range from 35 to 50 feet tall. Large white panicles of blooms appear in early spring, often covering the tree. Its leaves are palmate, or arranged in a fan shape around the petiole. These trees take on a raggedy appearance by summer's end, affected by a leaf blotch disease and, sometimes, by leaf scorch. Buckeyes don't transplant easily as they mature. However, their seeds are fairly easy to start. Sow them outside in the fall, but protect them from squirrels. If all goes well, they'll grow in the spring.

Pawpaw, Custard Apple (Asimina triloba)

This native plant is found along the stately cliffs of the cemetery's Plum Street entrance. Its large, floppy dog ear-sized leaves make it easy to spot. They're 9-10 inches long and turn a magnificent yellow. Its spring blooms become an incredible, velvety maroon, but are attached directly to the stems and can be quite difficult to spot for the untrained observer. Pawpaw fruits develop during the fall and are quite attractive to animals ... even to some humans. They bear the texture of ripe bananas, and some can't get past this to enjoy the fruits!

River Birch (Betula nigra)

River birch is native to some southeastern Ohio counties but grows well in Clark County. It's easily recognized by the curly, pinkish-tan, salmon-colored bark that peels from the branches of its young trunks. Older trunks appear a darker red to grayish brown as they exfoliate. This tree does quite well in wet soils, making it suitable for areas that might periodically flood. The river birch rises 60-70 feet tall and is generally a multi-stemmed tree. Other birch species don't grow well in the Springfield area, making this one a preferred selection.

American Hornbeam, Ironwood, Musclewood (Carpinus caroliniana)

The common names of musclewood and ironwood aptly describe the bark of this tree—a smooth, dark, bluish-gray that resembles taught, sinewy muscles. Native to Ohio and commonly distributed, this species thrives in a woodland setting, as an understory tree, or in a mid-sized landscape. Its height can rise to 30 feet, and at times, nearly as wide given its somewhat rounded growth habit. Some nature lovers inappropriately identify this tree as a beech (same family, different species).

Hickory (Carya spp.)

Ferncliff Cemetery is home to four species of hickory. When described, hickories are generally treated as a group. They possess limited landscape value and some are considered 'messy.' They're somewhat difficult to transplant, given their extensive taproot, and can reach 60 feet in height. Hickories display a wonderful, golden yellow fall color and, of course, their nuts are hunted by both animals and humans. Of Ferncliff's four species, shagbark is the most prominent and features a shaggy-bark appearance. Bruising or crushing its leaves unleashes a subtle apple scent, and its nuts are edible and sweet. The popular and aptly named shagbark is often used for hickory chips and smoking meats. The mockernut and the shellbark also produce nuts that are edible and sweet, while the pignut tends to bear bitter, astringent nuts.

Flowering Dogwood (Cornus florida)

An understory tree, flowering dogwoods can be found tucked in and among the cemetery's larger trees—truly their ideal location. When planted in full sun, poorly drained or harsh, dry soils, flowering dogwoods tend to decline. They're native from Massachusetts to Florida and west to Ontario, Texas and Mexico. Their beautiful flowers (which aren't actually flowers) are what make this tree attractive in a landscape. A flowering dogwood's true flowers are inconspicuous, greenish in color, and found in the center of the white bracts. The bracts appear from mid-April to May and show off in the cemetery before many of its neighbors leaf out. The dogwood's fruits, which turn red in the fall, are also quite showy. If planting a dogwood in an Ohio landscape, make sure its seed source is one hardy to Ohio. Many times, dogwoods from southern seed sources fail to bloom due to insufficient flower bud hardiness.

Kousa Dogwood (Cornus kousa)

A little-known dogwood that rivals the flowering version for flower power and fruiting interest, the kousa is an excellent choice for Ohio landscapes. This dogwood blooms a bit later into May and tends to have larger bracts. Its flowers are on little stocks that hold them slightly above the foliage, making them even showier. The fruits turn a beautiful red in the fall and resemble a large raspberry perched on a stem, creating a dramatic show during heavy fruiting years. The strong, horizontal lines of its branches also make this tree an excellent landscape choice to break up a vertical space.

Corneliancherry Dogwood (Cornus mas)

This tree is among the first to bloom inside Ferncliff each spring and may go unnoticed unless one visits in March. Its somewhat dull, yellow flowers cover the entire tree and are easily spotted without early competition from foliage or other flowers at bloom time. It tends to grow naturally as a small shrub (20-25 feet tall), but can be trained into a tree form.

European Filbert, Cobnut (Corylus avellana)

Tucked into a corner along McCreight Avenue, just opposite of Section M, is a European filbert planted by past superintendent Lou Shelton. This filbert is native to Europe, western Asia, and northern Africa and is prized for its nuts in European countries. The filbert can grow to 20 feet and usually forms a dense, rounded thicket of branches.

Hardy Rubber Tree (Eucommia ulmoides)

This tree isn't used much as a landscape item but has great potential. Its foliage displays a rough, course texture and remains green all summer. Its spring flowers are inconspicuous and its fall color fairly non-existent, but it suffers very few pest and disease problems and offers usefulness as a medium-sized landscape choice. Slowly tearing one of its leaves reveals latex-like strands that give this tree its common name.

Beech (Fagus spp.)

F. grandifolia American beech and Fagus sylvatica European beech are scattered throughout Ferncliff grounds. Clark County was once a beech/maple forest, and these shallow-rooted trees are frequently found in the area. The American beech is native and can also be found naturally occurring in county woodlands. Both trees possess similar growth habits, soaring to 50 or 60 feet, and display a fall show of golden bronze. American beech leaves are serrated, while European beech leaves are smooth and wavy along the margins. The bark of a European beech is generally darker and develops an elephant-hide texture and appearance. European beech branches also tend to grow closer to the ground. Of the European beech's numerous cultivars, three are found inside Ferncliff: Purpurea (purple leaves), Rohanii (brownish-purple leaves with undulating margins), and Tricolor (a purple leaf with a rose and pinkish-white border most beautiful during the early spring and summer).

Gingko, Maidenhair Tree (Gingko biloba)

Several gingkos are found within the cemetery; however, the largest and nicest one grows along the north side of the Plum Street entrance. The gingko hails from eastern China and is believed among the oldest species of trees. Its leaves are fan-shaped and grow in clusters of three to five atop spurs along its branches. The gingko provides significant texture in a landscape. It can grow quite large (up to 80 feet tall), and features a male and female version—of considerable importance when planting a gingko. Female trees produce fruits that are, simply put, smelly. In fact, their odor can be downright offensive. However, gingko seeds have a nice taste and are readily eaten in China and Japan. Many gardeners and landscape artists tend to plant male cultivars to avoid an unpleasant smell. The gingko is nonetheless a durable landscape tree, its fall color a beautiful, crisp yellow. It's easy to miss its colorful display, as its leaves tend to turn and drop quickly.

European or Common Larch (Larix decidua)

This tree might appear to be an evergreen to the untrained observer. Its leaves are needle-like, resembling those of the more well-known evergreen, but its leaves turn a wonderful yellow and drop in the fall, making it a deciduous tree. The larch grows quite large (to 75 feet). A beautiful specimen towers over the south side of Kelly Lake, a fitting place for this large, stately tree whose branches become more horizontal and droop with age. As a 'conifer' it bears cones. It's native to Europe and was introduced to the United States during colonial times.

Cucumbertree Magnolia (Magnolia acuminate)

Several common species of magnolia reside in Ferncliff, but the cemetery also features a particularly interesting species not often seen in

Clark County. They're planted in clusters of three near Sections 50 and 55, alongside the road from the Plum Street entrance. Capable of reaching 50 or 60 feet high and 20 to 30 feet wide, this tree's low branching structure makes it appear to be a large shrub. Its leaves grow to 10 inches long and are quite striking—a beautiful green during the summer with little fall color. Its handsome flowers appear from May to early June, but aren't always seen because they are borne near the tree's top. Its fruits are quite attractive, opening as fall approaches to reveal pinkish-red seeds. They don't last long but are certainly ornamental when they first appear. The cucumbertree magnolia is native to Ohio and commonly found in the eastern portion of the state.

Dawn Redwood
(Metasequoia glyptostroboides)

Another deciduous conifer, the dawn redwood is a magnificent specimen tree in the cemetery. It can grow to 80 feet and sports a pyramidal growth habit. Its soft, needle-like leaves are about 1-2 inches long and turn a delightful orange brown and russet red before dropping each fall. This redwood performs best in well-drained but moist soils and even tolerates standing water. Its reddish-brown bark tends to peel in thin strips as it ages. The dawn redwood is native to China but is becoming more readily planted in Ohio.

Black-gum, Sour-gum, Tupelo
(Nyssa sylvatica)

The tupelo is generally found on soils with a low pH (5.5-6.5), but a few of them grow along Ferncliff's gentle slopes. This tree is native to the eastern portion of the U.S. and appears in all parts of Ohio except the driest counties in the northwestern part of the state. It's considered an under-story tree or found along the edge of forests and wooded areas. Its glossy green foliage turns a beautiful blend of yellow, purple, red, orange and scarlet each fall, and the bark of mature tupelos can be quite striking, turning a medium gray with distinctly flat-topped blocks separated by deep crevices. Older bark's appearance becomes almost alligator-like. Its native habit ranges from Maine and Lake Erie's Ontario shore to Texas and mid-Florida, common among the hills of these areas. A tupelo's fruit ripens to bluish-black in September and October and is enjoyed by birds and mammals. The tree is difficult to transplant due to a long taproot.

American Hophornbeam, Ironwood
(Ostrya virginiana)

Although an excellent, medium-sized landscape grower, this tree isn't readily found in the nursery trade. Interestingly, its foliage resembles that of an elm or birch tree. It tends to grow in a rounded outline and features graceful, drooping branches. Its yellow fall color isn't highly noticeable, but once this tree is established it suffers very few problems, growing in wet or dry soils. Its Latin name comes from the Greek word 'ostrua,' a tree with very hard wood. Its more common name is derived from its fruit, which resembles that of a hops plant.

Swiss Stone Pine (Pinus cembra)

This pine is native to the mountains of central Europe and southern Asia and is one of only a few in Clark County. Although an excellent specimen tree, it's somewhat slow-growing (typically 30-40 feet tall and 15-25 feet wide). Ferncliff's Swiss stone pine is just 10 feet wide and perfectly sited on the northwest corner of Section F.

London Planetree (Platanus xacerifolia)

Numerous London planetrees are sprinkled throughout Ferncliff, with many planted in sections nearer the Plum Street side. This tree's beauty lies in its creamy to olive-green, exfoliating bark. A large, stately tree that can rise to 100 feet, it should be awarded plenty of room in a landscape. The planetree's fruits are rather large and occasionally messy. It's widely planted throughout London, England.

Sycamore, American Planetree (Platanus occidentalis)

A primary difference between the American and London planetree is that the American version's bark ranges from white to creamy white, as opposed to olive. A grand tree with broad-reaching branches that grow horizontally from the main trunk, this species grows in the center of the cemetery's GAR section.

Oak (Quercus spp.)

Ferncliff displays numerous oak species, particularly in a section near the Plum Street entrance. The most common species found are white, red, and scarlet. Oaks are typically divided into three subgroups, and Ferncliff boasts two of the three represented in Ohio: red or black oaks, and white oaks. The acorns are seldom used for human consumption, but provide an important food source for animals. Leaves of the red oak group feature bristles at the tips of their lobes and include red, scarlet, pin, willow and shingle. Leaves of the white oak group feature lobes without bristle tips and include white, English, burr, chinkapin, and swamp white oak.

White Oak (Q. alba)

White oak grows 50 to 80 feet tall and becomes upright to broad-rounded as it ages, creating an imposing specimen at maturity. Its bark ages to a light, ashy gray and is often broken into small, vertically arranged blocks and scales, becoming deeply fissured or irregularly plated with occasional smooth, gray spots. Fall color is a rich, red wine or brown, depending upon the seed source. White oaks are a large, valuable forest tree in the eastern United States. Their wood is often used in flooring, boat-building and wine and whiskey casks. They're most likely found in all of Ohio's counties.

Scarlet Oak (Q. coccinea)

Scarlet oaks reach 70 to 75 feet and display a glossy-green, summer foliage that turns scarlet and russet red during the fall. Its leaves are similar in shape to the red oak.

Shingle Oak (Q. imbricaria)

Reaching heights of 50 to 60 feet, its leaves don't resemble a typical oak leaf because they are long and slender instead of lobed. Shingle oaks' fall color ranges from yellow-brown to russet-red.

As its name implies, wood from this tree is used to make shingles.

Burr Oak (Q. macrocarpa)

These large trees fit perfectly into any haunted house scene. Their coarse, massive form and broad crown of stout branches create the perfect spooky backdrop. The burr oak soars 80 feet in height and has an equally or slightly greater spread. Its bark is quite rough and develops deep ridges and furrows with age. The burr oak was once an important constituent of the Great Black Swamp area in northwestern Ohio and in oak-hickory areas of west-central Ohio. Interestingly, a cap covers most of its acorn and features a fringed margin. Burr oaks are highly tolerant of dry, clay soils and city conditions, more so than most oaks, but do not transplant easily. Their fall wardrobe features a dull, washed-out yellow.

Chinkapin Oak (Q. muehlenbergii)

Although not found very often in Clark County, a chinkapin oak can be found in the cemetery. Quite a few of these reside in the Bellefontaine area and are also found in southwestern Ohio's limestone soils. This tree tolerates dry, rocky conditions, growing to roughly 50 feet under landscape conditions, 80 feet in the wild. Its spread is normally greater than its height. The chinkapin oak develops an open, rounded crown at maturity and possesses a medium to slow growth rate. It's difficult to transplant, with leaves not typical of oak trees—narrow, 4 to 6.5 inches long, with coarsely tooted edges.

Pin Oak (Q. palustris)

A pin oak becomes a beautiful tree when grown in the right soil pH. Unfortunately, Clark County's pH level is roughly 7.0 or higher and not conducive to their growth. As of 2008, only one pin oak stood near the cemetery's entrance, growing but not thriving. Generally, during midsummer the pin oak exhibits chlorosis (a yellowing of its leaves). It can reach 70 feet and boasts a strong, pyramidal habit. Its branches tend to be pendulous around the tree's bottom, horizontal throughout the middle, and upright in its higher areas, resulting in its distinctive, regal shape. Pin oaks prefer an acidic soil (between 6.0 and 6.5 for the best growth), are easily transplanted, and make nice street trees because of their shape.

Willow Oak (Q. phellos)

Willow oaks don't normally survive in central Ohio, tending to do better in southern Ohio and beyond, and is not readily hardy in the Clark County area. However, Ferncliff features one toward the back of its grounds, near the river. These trees can grow to 60 feet tall and 40 feet wide, developing either dense, oblong-oval or rounded crowns as they mature. Their leaves are quite narrow (less than an inch wide), average 3 to 5 inches long, and resemble willow leaves. In the fall they range from shades of yellow, bronze or orange to yellow-browns and russet reds. Their willowy appearance adds nice texture to any landscape. Their acorns are quite tiny—less than a half-inch.

Chestnut Oak (Q. prinus)

Stretching 60 to 70 feet tall, its leaves are 4-6 inches long with slightly toothed margins. Their fall colors range from orange-yellow to yellow, and their acorns are relished by many types of wildlife.

English Oak (Q. robur)

The upright variety is very columnar in habit and can reach 15 feet wide and 50 feet tall.

Red or Northern Red Oak (Q. rubra)

The red oak is one of the most common oak species in the cemetery. Its leaves are 4-8 inches long and lobed with pointed tips. This tree can reach 75 feet and boasts a fall color of russet to bright red, although some seedling varieties may not color at all. Oaks cross-pollinate and hybridize freely, making it somewhat difficult at times to identify.

Common Sassafras (Sassafras albidum)

This native Ohio plant has three different leaf forms on the same plant. Leaves are either entire or simple, one-lobed or mitten-shaped, or three-lobed and boast outstanding fall color, changing from yellows and oranges to scarlets and purples. Sassafras is common to woodland areas and is a nice, naturalized plant. It's sometimes difficult to get started in the Clark County area and should be planted during early spring, as it has a deep taproot. The bark of its roots is used for tea; the oil distilled from its roots is used to perfume soap.

Japanese Pagodatree (Sophora japonica)

This is one of the latest-blooming trees in the cemetery during the summer. Its white panicles are mildly fragrant and can grow as large as 10 inches in length. The Japanese pagodatree thrives in poor soil conditions, can reach 60 feet in height and requires a large spread area. It's best placed in open areas due to its messy nature, with the dropping of petals and subsequent fruits, and offers very little fall color.

Common Baldcypress (Taxodium distichum)

Another stately beauty inside Ferncliff Cemetery, the baldcypress lines the banks of Kelly Lake, gracing the area with delicate foliage and interesting cypress knees. Its delicate, evergreen-like leaves are similar to the dawn redwood, except they are whorled along their stems, and turn a beautiful russet brown to orange bronze before dropping. An excellent tree for wet areas—if planted in a moist or wet area, cypress knees develop. These knees are protrusions, similar in color to the trunk of the tree, and rise from the ground from 1 to 8 inches. The landscape arrangement of baldcypress and larch trees around the pond creates a gorgeous setting for artists and photographers.

Japanese Zelkova (Zelkova serrata)

This tree was once thought capable of replacing the American elm as a street tree. Its upright, arching growth habit makes it perfect for the job, but several years it was introduced into the industry, it was found to have a disease problem that precluded it from becoming a popular street tree. Several are planted throughout the cemetery and it's easy to spot their vase-shaped growth habit. Growing 50 or 60 feet tall with an attractive, smooth, gray bark, they develop a slight, brownish-green exfoliating character and are mostly found in newer sections of Ferncliff.

Other, less prominent Ferncliff trees include: Briotii Red Horsechestnut, Boxelder, Ash-leaved Maple, Japanese Maple, Norway Maple, Red Maple, Silver Maple, European Hornbeam, Northern Catalpa, Hackberry, Eastern Redbud, American Yellowwood, White Fringetree, Cockspur Hawthorn, Washington Hawthorn, Thicket Hawthorn, Shadblow Serviceberry, Black Walnut, White Ash, Green Ash, Blue Ash, Thornless Common Honeylocust, Kentucky Coffeetree, American Sweetgum, Osage-orange, Hedge-apple, Saucer Magnolia, Star Magnolia, Crabapple, Common Mulberry, Sourwood, Norway Spruce, White Spruce, Austrian Pine, Colorado Blue Spruce, Hoopsii, Eastern White Pine, Scotch Pine, Eastern Cottonwood, Black Cherry, Japanese Kwanzan Flowering Cherry, Weeping Higan Cherry, Shubert Common Chokecherry, Yoshino Cherry, Callery Pear, Black Locust, Weeping Willow, Japanese Tree Lilac, American Linden, Littleleaf Linden, Crimean Linden, Silver Linden, Canadian Hemlock and Slippery Elm.

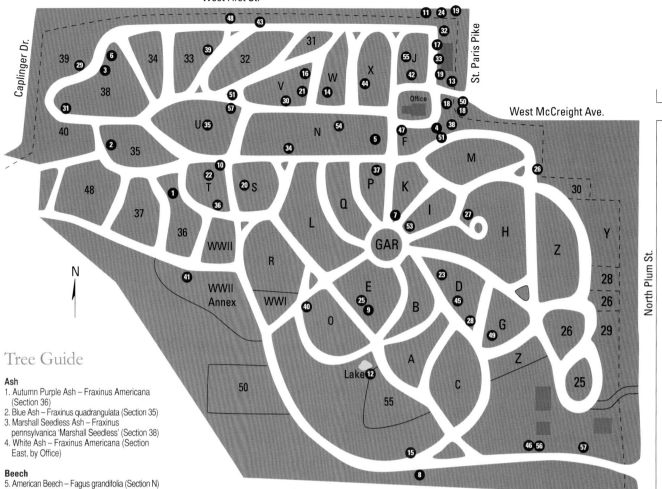

Tree Guide

Ash
1. Autumn Purple Ash – Fraxinus Americana (Section 36)
2. Blue Ash – Fraxinus quadrangulata (Section 35)
3. Marshall Seedless Ash – Fraxinus pennsylvanica 'Marshall Seedless' (Section 38)
4. White Ash – Fraxinus Americana (Section East, by Office)

Beech
5. American Beech – Fagus grandifolia (Section N)
6. European Tricolor Beech – Fagus sylvatica (Section 38)

Black Gum "Tupelo"
7. Black Tupelo – Nyssa sylvatica (Section K)

Box Elder Maple
8. Box Elder – Acer negundo (Lower Road Area)

Buckeye
9. Ohio Buckeye – Aesculus giabra (Section E)

Cherry
10. Weeping Higan Cherry – Prunus subhirtella (Section U)

Crabapple
11. Malus – (Section W)

Cypress
12. Bald Cypress – Taxodium distichum (Section Pond Area)

Dogwood
13. Cornelian Cherry Dogwood – Cornus mas (Mausoleum)
14. Flowering Dogwood – Cornus florida (Section W)

Elm
15. Slippery Elm – Ulmus rubra (Lower Road Area) Ohio Champion 07-06-03

Euonymus
16. Winged Euonymus – Euonymus winged (Section V)

Fir
17. White Fir – Ables concolor (Mausoleum)
18. Douglas Fir – Pseudotsuga menziesii (Main Gate)

Fringe Tree
19. White Fringe Tree – Chionanthus virginicus (NE Corner)

Ginko
20. Ginkgo Tree – Ginkgo biloba (Section S)

Hackberry
21. Common Hackberry – Celtis occidentalis (Section V)

Hawthorne
22. Downy Hawthorne – Crataegus mollis (Section T) Ohio Champion 02-28-04
23. Washington Hawthorne – Crataegus phaenopyrum (Section D)

Hemlock
24. Canadian Hemlock – Tsuga Canadensis (First Street)

Hickory
25. Shagbark Hickory – Carya ovata (Section E)

Hornbeam
26. European Upright Hornbeam – Carpinus betulus "columnaris" (By Section 30)

Horsechestnut
27. Common Horsechestnut – Aesculus hippocastanum (Section H)
28. Red Horsechestnut – Aesculus carnea (Section D)

Japanese Pagoda Tree
29. Japanese Pagoda – Sophora japonica (First Street)

Linden
30. American Linden – Tilia Americana (Section V)
31. Silver Linden – Tilia tomentosa (Section 38)
32. Greenspire Linden – Tilia cordata (Mausloeum)

Maple
33. Hedge Maple – Acer campestre (Mausloeum)
34. Japanese Red Maple – Acer palmatum (Section N)
35. Norway Maple – Acer platanoides (Section U)
36. Red Maple – Acer rubrum (Section T)
37. Red Sunset Maple – Acer rubrum (Section P)
38. Silver Maple – Acer saccharinum (Office)
39. Sugar Maple – Acer saccharum (Section 33)

Oak
40. Bur Oak – Quercus macrocarpa (Section O)
41. English Oak – Quercus robur (Section 36)
42. Pin Oak – Quercus palustris (Section J)
43. Red Oak – Quercus rubra (First Street)
44. Scarlet Oak – Quercus coccinea (Section X)
45. White Oak – Quercus alba (Section D)

Pawpaw
46. Common Pawpaw – Asimina trioba (Lower Road)

Pine
47. Swiss Stone Pine – Pinus cambra (Office)
48. Scotch Pine – Pinus sylvestris (First Street)

Redbud
49. Eastern Redbud – Cercis Canadensis (Section C)

Rubber Tree
50. Hardy Rubber Tree – Eucommia ulmoides (Mausoleum)

Sassafras
51. Common Sassafras – Sassafras albidum (Section F)

Spruce
52. Colorado Blue Spruce – Picea pungens v. glauca (Triangle by Section U and V)

Viburnum
53. Blackhaw Viburnum – Viburnum prunifolium (Section I) Ohio Champion 07-14-04
54. Rusty Blackhaw Viburnum – Viburnum rufidulum (Section N) Ohio Champion 06-07-04

Zelkova
55. Japanese Zelkova – Zelkova serrata (Section J)

Other Sites
56. Widow's Walk – Carved from the stone between the cliffs, this stairway was made as a short cut for those visitors who came on foot to visit the cemetery.
57. Leaning Rock – Has many stories about it including the Indian Lore of Weeping Rock as well as a possible grave site for workers from the National Road Project, also used as their campsite.

Biographies
NOTABLE GRAVESITES

Captain Richard Bacon
Anner Fosdick
(February 7, 1757 - November 2, 1822)
(1761 - August 29, 1821)

Richard Bacon was born in Middleton, Connecticut, the son of Zaccheus and Mercy Hubbard Bacon. The details of his early life are missing until his enlistment in military service.

According to the *Records of Connecticut Men in the Revolutionary War*, Bacon was a private in Captain Wells' Company, Colonel Erastus Wolcott Regiment, Connecticut Troops 1776, a private in Captain Wells' Company, Colonel S.B. Webb's Regiment, Connecticut Troops 1777, and a private in Captain Hopkins' Company, Colonel S.B. Webb's Regiment, 1781. The *Daughters of the American Revolution Patriot List* refers to him as Sergeant Bacon.

Bacon married Anner Fosdick of Wetherfield, Connecticut on December 26, 1784 and the couple produced seven children: Richard, Henry, George, Charles, Allyn, John and Samuel. John Bacon, a prominent figure in the banking business, was born October 8, 1797 and died March 5, 1870. He married Mary Ann Cavileer on March 1, 1820, and at the time of his death, served as director and president of Mad River National Bank.

In later years Richard Bacon moved to Dayton and lived with son Henry, a practicing attorney and one-time Urbana resident. He died on November 2, 1822 in Dayton.

When Ferncliff Cemetery was established, Captain Richard and Anner Bacon were transferred from Dayton to be buried in Section E, Lots 19 and 20, alongside family members.

Jonathan (Mulholland) Milhollin
(1765 - 1834)

Little is known regarding the early life of Jonathan Milhollin, born in Loudon County, Virginia. In 1781 he served as a private in the Revolutionary War, a draftee militiaman for two months and for two more months, a substitute. That same year he also enlisted and served for 15 months under Colonel Amond and was discharged in 1783.

After the war Milhollin married Nancy McClintic, daughter of William and Nancy Shanklin McClintic, and started a family. The family resided in Virginia until a 1788 move to Kentucky and, in 1800, a trip to Ohio, where Milhollin operated a grist mill and distillery.

Jonathan Milhollin died in October of 1834 at the age of 69 and was interred in the Milhollin burial grounds. His will, probated October 20, 1834, mentions his sons (McClintock, William and then-deceased Jonathan) and daughters (Nancy, Sarah, Jean Elizabeth, Mary, Jane and Margaret), as well as his grand-children.

Beginning March 18, 1868 their bodies were disinterred from the Milhollin family burial grounds and placed in Ferncliff Cemetery, Section D, Lot 107.

Maddox Fisher
(August 17, 1771 - October 22, 1836)

A native of Delaware state, Maddox Fisher lived with his family until manhood and married in 1791. He immigrated to Lexington, Kentucky at the turn of the century, where he amassed considerable wealth through manufacturing and real estate. He and his family migrated to Springfield in 1813 and quickly invested in the village of Springfield, purchasing from founder James Demint 25 lots at $25 each, most located near the public square. Being a man of significant business ability, Fisher erected several mills, including a sawmill and a cotton mill on the banks of Buck Creek. Several years of prosperity led to a flour mill and a dry goods store on Main Street.

His wealth and business acumen helped the town flourish as he built numerous residential and retail establishments. He constructed the Fisher Building on the corner of Main and Limestone streets, living on the second floor with his family while utilizing the ground floor as a store.

In 1818, with the state legislature in Chillicothe to debate construction of a new county from adjoining counties Madison, Greene and Champaign, Fisher rode his horse to the city site and successfully lobbied a bill that established Clark County, with Springfield as its county seat. There had been debate as to whether Springfield or New Boston, reputed birthplace of Tecumseh, should be selected as county seat, but Fisher's opposition to New Boston tipped the decision to Springfield. In gratitude, he was named Springfield's postmaster.

Maddox Fisher and his wife, Mary, were members of Methodist Episcopal Church. He was elected township trustee in 1819 and Overseer of the Poor in 1827, remaining active within the community until his falling ill in 1835. An *Ohio Pioneer and Clark County Advertiser* obituary marked his death on October 22, 1836 at the age of 66: "Brother Fisher was a native of Delaware, immigrated to this Western Valley and lived several years in the State of Kentucky. He was confined by a stroke of palsy for one year."

Mary Fisher died on October 17, 1838 in her 67th year. Both were interred in Columbia Street's Demint Cemetery, but later transferred to Ferncliff in 1875. Named in Fisher's will, written March 5, 1836, were his wife, Mary, daughters Nancy N. Fisher Perrin, Eliza Fisher Carter and Rebecca Fisher Baird, sons William L. Fisher, John Fisher and Maddox Fisher Jr., and several grandchildren.

Maddox Fisher Jr. was born in Lexington, Kentucky on November 29, 1812 and was just one year old when his family arrived in Springfield. He married Sarah T. Barrett, daughter of Captain Abner Barrett of Champaign County, Ohio. The couple had one child, Charles B. Fisher. After his dad's death, Maddox assumed all business operations.

He remained active in public affairs, just as his father before him. He was among Ferncliff Cemetery's initial trustees and first 33 subscribers. He died on November 6, 1889, a few days short of his 77th year, inside his residence on the corner of Limestone and Main streets. At the time of his death, Maddox Fisher Jr. was Springfield's oldest citizen. Mrs. Fisher lived until 1904. They rest in Section A, Lot 14 among Fisher and Barrett family members.

Henry Bechtle
Elizabeth Perry
(May 27, 1782 - February 9, 1839)
(1786 - September 6, 1869)

The first Bechtle family member to arrive in America was Jacob, who emigrated from Manheim, Germany while the U.S. was in its infancy. He settled in Pennsylvania and took up farming along the banks of the Schuylkill River, married and reared a family. The migration began in the next generation, when the families moved south before settling in Hagerstown, Maryland.

Henry Bechtle was born a few miles from Hagerstown on May 27, 1782, where he was reared and educated. In 1802 he joined General Bronson for a westward horseback tour through the wild country to Arkansas. One of his studies was civil engineering, which helped him when he assisted that state in laying out its counties. After several months he returned to Maryland, but was not content after seeing life on the western frontier and so embarked upon another journey that in 1804 took him to Cincinnati. There he established a flour and produce business, shipping his products on barges to New Orleans. After disposing of the freight and selling the barges, he bought a horse for his return trip to Cincinnati.

In 1816 Bechtle married Elizabeth Perry, the daughter of Captain David and Susan Perry and born near Lexington, Kentucky. Six of the couple's eight children were reared to maturity. At the time of their marriage Bechtle worked in the wholesale and retail dry goods business in Cincinnati. He bought roughly 1,000 acres of Springfield land in 1818, but continued his Cincinnati-based business until 1826, when he moved to Springfield permanently and devoted his life's remainder to improving his land and mill operations. He amassed a large fortune, which included land in Illinois.

Henry Bechtle was selected juror for the year on October 14, 1830, becoming active in local politics and a staunch supporter of the Democratic Party. He was a great admirer of General Jackson, and many of the noted men of his day, including General Harrison and Henry Clay, were frequent guests in his home. Bechtle was known as a kind and hospitable gentleman.

Following his death on February 9, 1839, Bechtle's widow and children maintained his business enterprise for a number of years.

Because of the city's rapid growth, the Springfield Cemetery Association recognized the need for additional land for burials. On September 12, 1863 the Association purchased from Bechtle's widow and heirs 70.8 acres of ground northwest of the city on Plum Street, overlooking the hills and cliffs north of Buck Creek. The purchase price was $7,030. The name "Ferncliff" was immediately adopted for the new cemetery grounds, due to the rugged, riverside beauty of the area.

Elizabeth Bechtle died on September 6, 1869 at the age of 82. She rests in peace with her husband and children in Section C, Lot 20, each of them having been moved to Ferncliff from Columbia Street's Demint Cemetery on June 26, 1865.

Robert Layne
(July 11, 1757 - April 12, 1845)

Robert Layne, the son of Ayres and Mary Woodson Layne, was born July 11, 1758 on the Cattails of Eastern Licking Hole Creek (Goochland County), Virginia. Nothing is known about his early life except that he could write and that he joined the Goochland County Militia.

Layne served in the American Revolution in 1779 as a private in Goochland County. His company was commanded by Captains Jolly Parish and Nathaniel Massie. He later moved from Goochland to Albemarle County, where he married and remained for some 20 years after the war. His final two moves were to Kenawha and Clark County, Ohio, respectively.

W.H. Harris related in a paper delivered to a pioneer meeting held near Emery Chapel, "In the first house north of Possum Road lived old Mr. Lane [sic]. All knew him as 'Granddaddy Lane (sic).' He was a soldier in the Revolution and the only one I ever saw. The boys from Possum School used to go to his place for apples. The trees had grown from seed and the apples were sweet."

Robert died on April 12, 1845 in Clark County. He and his family were later removed from the family burial grounds to Ferncliff Cemetery, Section L, Lot 367.

Ezra Keller, D.D.
Caroline Routzahn
(May 12, 1812 - December 29, 1848)
(1819 - September 19, 1888)

Ezra Keller was the third son of six children born in Middleton (Frederick County), Maryland to Jacob and Rosanna Daub Keller. Baptized on June 21, 1812, his early years were spent performing routine chores on the family farm. Two of his brothers died in their youth.

Nearing the age of 12 Ezra was sent to a German school and taught by a devotedly pious man who instilled within him the importance of religious and devotional exercises. Similarly inspired by a young farm worker named William

Lingenfelter, who talked of God and religion with ease, Ezra soon accompanied him during his preachings. About the same time, he also earned the encouragement of Reverend A. Peck, who challenged the young Ezra to follow and apply his beliefs and attend Pennsylvania College to secure an education.

Keller left home to live with Rev. Peck and his family on March 20, 1830, excited to further his studies. He received catechistical instruction and was confirmed to the church. So determined was he to extend his education that he walked 40 miles from Middletown to Gettysburg, Pennsylvania, arriving April 17, 1830 to begin his Preparatory Class studies.

Ezra graduated from Pennsylvania College at Gettysburg (today known as Gettysburg College) in the fall of 1835 and began missionary work at Taneytown and Hagerstown, Maryland, later traveling as far as Alton, Illinois. He proposed to Caroline Routzahn in 1835 and the couple married on April 25, 1837. Six children were born, but three preceded Ezra in death.

Keller's final missionary tour brought him to Clark County, Ohio, where he opened a grammar school on November 3, 1845 inside the rooms of First English Evangelical Lutheran Church. Once Wittenberg College was chartered in 1845, Dr. Keller became the institution's first president. Progress continued on the school until his sudden death on December 29, 1848 at the age of 37.

The Tri-Weekly Republic of December 30, 1848 noted in Dr. Keller's obituary: "We are pained to announce that Ezra Keller, D.D., of Wittenberg College is no more. He died in his residence in this town yesterday morning at one o'clock of congestive or brain fever. Although his degree was one usually indicating age, Dr. Keller was quite a young man—we think but now yet 40.

"He was a ripe scholar, an exemplary Christian, a patriot and a philanthropist, and gave promise of a brilliant reputation. As a preacher few could exceed him. His language was earnest, expressive and simple in the extreme. A child could understand him, but all felt the power of his words. The community and college will both suffer by his loss—time alone can tell how deeply. The funeral will take place at 11 o'clock this morning from his residence."

Dr. Ezra Keller was buried in Woodshade Cemetery, but later transferred to Ferncliff on November 18, 1868 to rest among family in Section E, Lot 25.

The Biography of Ezra Keller, D.D. was published in 1859 by Rev. M. Diehl, A.M., a former professor of ancient languages at Wittenberg College. Rev. S. Sprecher, D.D., onetime Wittenberg president, penned the book's introduction. Material for the work was gathered from Keller's diary and assorted documents found inside one of his old trunks. Among them were Dr. Keller's Pennsylvania College diploma, dated September 18, 1835, and certificates of membership in the Phrenokosmian Literary Society and the Temperance Society of the State of Maryland. Certificates of licensure and ordination were also found.

Jeremiah Warder
Ann Aston
(1780-1849)
(1781-1871)

Jeremiah Warder, son of John and Ann Head Warder, arrived from Pennsylvania in the early 1800s to oversee some of his father's extensive land holdings consisting of over 200 acres of land. He was impressed with what he saw and following his father's death in 1828 made the decision to leave their home in Philadelphia and move west to the frontier. Jeremiah, Ann and some of their children arrived in Springfield on June 1830.

By 1830 Jeremiah purchased from his father's estate the village of Lagonda consisting of several dwellings, sawmills, a woolen factory and gristmill, then proceeded to make improvements by erecting a large mill and building a dam over the stream to increase water power for more mills. One of the early mills operated by Jeremiah was the Warder-McLaughlin Woolen Mill. In 1831 he purchased the *Western Pioneer* newspaper later turning it over to Edward Cummings as editor and John M Gallagher. Edward Cummings later married Sarah Warder, daughter of Jeremiah and Ann, on February, 1833. Jeremiah was quite active in community affairs until his death in 1849, having voted in 1833 in the General Assembly to incorporate Springfield High School into the city of Springfield, elected Overseer of the Poor in 1833 with Griffith Foos, and corresponding secretary of the Clark County Agricultural Society.

To provide proper training for her children as well as others Ann opened a school in her home and later opened another school for the more advanced students. She was an early active leader in the support of the Springfield Underground Railroad movement, and a generous contributor to all charitable causes. Her tremendous drive for the betterment of her beloved town continued until her death on August 10, 1871 at the age of 86 years.

The family continued interests in Warder, Mitchell and Company, later known as Warder, Mitchell and Glessner, manufacturers of the Champion Reaper. The Company then became known as Warder Bushnell and Glessner Company manufacturer of Reapers, Movers and Self Binders, continuing their operation until the turn of the century, when the company merged with other companies and became known as International Harvester Company. Through these efforts a great fortune was amassed with much of the fortune being used for philanthropic activities, contributing greatly to these institutions.

William, third son of Jeremiah and Ann, first entered business in connection with the Lagonda Agricultural Works, later forming a lifetime partnership with William A. Barnett in the milling business. When the Springfield Board of Trade was organized in 1865 with a membership of 165 he was elected president. After several years the organization was abandoned. In 1863 at the council meeting he offered a resolution to organize an association to study the feasibility of another cemetery in Springfield due to the population growth. He was then appointed chairman of the council committee on cemeteries in 1863 which afterward resulted in the formation of the Ferncliff Cemetery Association. About 1855 he married Miss Mary Price of Philadelphia with four children born to them.

He became incapacitated following a stroke of paralysis, but lingered on for several years until his death on August 31, 1886.

James Thompson Warder was affiliated with the firm of Warder, Brokaw and Child, the original Champion firm at Lagonda, then devoted most of his life to agricultural pursuits on the farm homestead known as "Woodside." He married Mary Wheldon daughter of Joseph Wheldon on January 1, 1855 and then there were two sons, Wheldon and Frank. He became ill from Bright's disease and succumbed on June 18, 1891 in his sixty-ninth year. His wife, Mary, followed him in death six month later on December 11, 1891.

Benjamin Head Warder, the youngest son, received his education in the Springfield seminary, later attending Cincinnati College, after which he read law for a short time. He then entered the family business of operating the gristmills and sawmills at Lagonda, later forming partnerships and mergers for the manufacture of agricultural implements to which he devoted most of his life.

During the Civil War, he was listed as First Lieutenant in the Ohio Militia, 35th Battalion, Company F from Lagonda of 73 men in 1863, and First Lieutenant 152nd Ohio Volunteer Infantry, Company F. The regiment saw service in Virginia.

In December, 1867, he married Miss Ellen N. Ormsbee, daughter of Benjamin Ormsbee of this city, with four children born to them.

Following the history of the benevolence of the Warder family, Benjamin became involved in

Benjamin Warder

efforts to establish a free library for the benefit of the citizens of Springfield. The family donated the land and money for that purpose to be erected in the memory of Jeremiah Anne Warder, thus the Warder Public Free Library came into being with the building formally dedicated on June 12, 1890.

In 1884 the family moved permanently to Washington where he became interested in real estate, becoming a stockholder in several banks, and director and stockholder of the American Security and Trust Company. He remained stockholder and president of the Springfield First National Bank.

While on an extended trip with the family, sailing from New York November 30, 1893 to Naples, Italy then to Egypt, Benjamin became ill while in Alexandria, dying on January, 1894. All of his siblings pre-deceased him except Mary Aston who married Charles Rannells and resided in St. Louis at his death.

Many members of this kind and benevolent family rest together in Section C of Ferncliff.

Charles Turner Cavileer
Rachel Trease
(1787 - October 15, 1850)
(1795 - September 6, 1887)

One of at least five children born to Charles and Ann Cavileer, Charles began life in Kent County, Maryland in 1787. He married Rachel (Treece) Trease of Northumberland, Pennsylvania on July 10, 1815, and shortly thereafter, the Cavileer family and siblings began a long journey westward, arriving in the village of Springfield, Ohio around 1818. He is listed in a Springfield Township Poll Book for May of that year.

From an early date Cavileer showed concern for the community's welfare, as his numerous contributions of both labor and money attest. He was elected Overseer of the Poor, served as a Springfield trustee and a Springfield Township juror, helped organize the Independent Fire Department on April 3, 1838, was elected Clark County Agricultural Society treasurer (1842-1843), donated money toward construction of the first county court house, and in December of 1835, was one of several individuals appointed to direct a newly erected county infirmary.

Charles and Edmund B. Cavileer were early stove and tinware business merchants. Clark County court records list them on November 21, 1820 as seeking renewal of a license to operate their business for another year at the cost of $15.00. Much later, the business was run by Charles' son, Alfred, and son-in-law, William McCuddy. Edmund B. died on October 3, 1849 at the age of 57. Charles' death followed on October 15, 1850. He was 63 years old. An October 18 obituary in the *Republic* described Charles Cavileer as "a resident of Springfield from an early day" and a "virtuous, useful and most excellent citizen" whose loss "is deeply felt by the whole community."

Charles and Rachel were blessed with at least 13 children, five of whom are known to have lived to adulthood. After her husband's death, Rachel continued life at the family homestead on the northwest corner of High and Spring streets until her death at age 92 on September 6, 1887. She and family members are interred in Ferncliff Cemetery, Section E. A few relatives' bodies were later transferred to Ferncliff from nearby Demint and Greenmount cemeteries.

Cavileer's son, Alfred, was born on January 30, 1825 in the same Main Street location where he later operated Cavileer & McCuddy, the stove and tinware business established by his father. Brother-in-law William McCuddy married Iserelda Gordon on December 19, 1848 and the couple had five children: Rachel, Charles, Mary C., Alfred, and Florence.

Alfred died of kidney disease on April 8, 1879 at the age of 54. The *Springfield Republic* described him as a "quiet man" and a "good citizen" who had been a "consistent member of the Second Presbyterian Church for 25 years."

Charles Turner, oldest son of Rachel and Charles, was born March 6, 1818 in the village of Springfield. Possessing an adventurous streak, he moved at age 17 to Mt. Carmel, Illinois, where he apprenticed as a saddle-maker for his uncle, Charles Constable. He eventually moved to Red Rock, near St. Paul, Minnesota, where he invested in real estate and operated a harness shop and drug store. Charles also studied medicine and collected medical books. He was appointed territorial librarian in 1849 by Governor Alexander Ramsey.

In 1851 a group of settlers sponsored by Charles Cavileer began the first permanent farming community at Pembina, a fir-trading station then located in the Minnesota Territory. He later became Pembina's postmaster and served in that capacity until near the time of his death on July 27, 1902.

In 1856 Charles wed Isabella Murray, daughter of Honorable Donald and Mary Jane Heron Murray, whose family hailed from the Red River settlement now known as Winnipeg. Isabella was born on August 7, 1839 in Manitoba, Canada. She died on February 14, 1920 and is buried with family in Pembina City, North Dakota. She was the mother of five children: Edmond, William McMurray, Albert Donald, Sara Jane, and Lulah Belle.

Cavileer County was eventually established in 1873 from the western part of Pembina County and named for Charles Cavileer, a well-known fur trader, by the Territorial Legislature. As was written about his father, Charles "was a virtuous, useful and most excellent citizen, and the loss is deeply felt by the whole community."

Pierson Spinning
Mary Schooley
(1786 - January 21, 1857)
(1790 - February 10, 1876)

Pierson Spinning's family was among the earliest of Clark County settlers, arriving in 1812 from nearby Dayton, Ohio. He was born in Elizabethtown, New Jersey, the son of Isaac and Catherine Pierson Spinning. He was still a young any when his family migrated to the Ohio territory mainly by flatboat, traveling the Ohio River to Cincinnati. Pierson Spinning began his career clerking in the dry goods store of H.G. Phillips, eventually going into business on his own in Middletown (Butler County), Ohio. He married Mary Schooley, daughter of John and Mary Earl Schooley, on September 9, 1812 in Springdale (Hamilton County), Ohio. The couple moved to Springfield and Pierson continued to operate his dry goods business—widely considered Springfield's first such operation. The couple had 11 children.

For many years it was Spinning's custom to travel by horseback to Philadelphia and New York City, where he'd purchase dry goods to be shipped via wagons over the Allegheny Mountains to Pittsburgh. From there, they'd travel by boat along the Ohio River to Cincinnati and again by wagon to their final destination in Springfield. He made the harsh trip despite being lame, riding side saddle to manage his limitation. Not surprisingly, due to such hard work and persistence, Pierson Spinning became quite successful, accumulating a large fortune that he invested in land. In 1832 while on one of his trips he arranged delivery of Springfield's first piano. Large crowds gathered whenever one of Spinning's daughters played, marveling at the music. The beautiful and historic instrument is today housed at the Clark County Historical Society.

Spinning constructed his East High Street residence in 1827 and, three years later, erected a large hotel building known as the Buckeye House, which he operated for a time. A contractor, he constructed part of the Miami and Erie Canal and a portion of the old National Road. He held several elected positions in Springfield, including justice of the peace, grand juror, commissioner and fence viewer.

A determined pioneer who contributed greatly to Springfield's prosperity through his accumulated wealth and energy, Pierson Spinning died of pneumonia on January 21, 1857 at the age of 70. In 1866, Mary Spinning moved to Dayton, Ohio and lived with her daughters until her death on February 10, 1876. Family members were buried in Demint and Greenmount cemeteries and later transferred to Ferncliff Cemetery, Section B, Lot 14.

John Humphrey
Jane Ward
(March 6, 1764 - March 19, 1857)
(December 10, 1771 - March 21, 1849)

Among Clark County's earliest settlers are members of the John Humphrey family, who joined Simon Kenton and James Demint in 1799. Six other Kentucky families soon followed. A fort consisting of 14 cabins within its pickets was erected near the Mad River west of town, offering settlers protection against the Indians. Humphrey later relocated to a tract of land northwest of Springfield, living there most of his life until the infirmities of advanced age forced him to sell his farm and move to the city.

Humphrey was born on March 6, 1764 in County Tyrone, Ireland. He immigrated to America in 1780 and settled in Greenbriar County, Virginia, where he married Jane Ward on November 25, 1790. The couple's first child, James, arrived on September 12, 1791. He soon had 13 siblings, eleven of whom grew to maturity.

The family moved westward into Mason County, Kentucky in 1793, and by 1799 arrived in Ohio and settled along the Mad River a short distance from town.

The Humphrey were staunch Presbyterians, so it was not unusual that John was elected an Elder of the Presbyterian Church on July 17, 1819. A venerable citizen, he remained active in the church until his death at age 93 on March 19, 1857 at a son's residence.

On March 27, 1857, *The Republic* wrote of his passing, "We barely noticed, last week, the decease of this aged patriarch, who departed full of years, and full of glorious hopes of those who have kept the faith and fought the good fight of the Great Sovereign—the kind maker and master of us all.

"Father (Humphrey) was one of the earliest settlers of this county—a native of the 'Emerald Isle'—and possessed such marked characteristics of head and heart as made his memory peculiarly fragrant in the minds of a large circle of kindred, and an almost innumerable multitude of friends, neighbors and acquaintances. He was a resolute Christian man throughout his life, and contemplated death with a most delightful composure, as it were the entrance upon a journey, the end of which, a land of peace, a home blessed and eternal."

John is buried in the Humphrey family burial ground with his wife and other family members. They were transferred to Ferncliff Cemetery to rest in Section E, Lot 103.

Griffith Foos
Elizabeth Whitsett
(1767 - 1859)

(1777 - 1833)

Nancy Winans
(1798 - 1858)

William Foos
(1814 - 1892)

Gustavus Foos
(1818 - 1900)

John Foos
(1826 - 1908)

Griffith Foos, son of John and second wife, Hannah Griffith, was born in Chester County, Pennsylvania in 1767. After the Revolution the family moved to Tennessee and, later, to Harrison County, Kentucky, where John married his third wife, Jane McCandless.

Joseph Foos, brother of Griffith, left the Kentucky regions in 1798 with the family seeking new territory to settle. They were among the first pioneers of Franklinton, now Columbus, Ohio, which had been plotted a year earlier by Lucas Sullivant. Foos established a Scioto River ferry operation he maintained until 1816. By 1803, he'd been granted a license to operate a tavern, an establishment that served travelers for a number of years. His father, John, and his wife, Jane McCandless, arrived in Franklinton about the same time.

Griffith and Elizabeth Foos joined other Kentuckians in 1800 and made a trip to Ohio. The group headed toward Franklinton but found the area's marshy area unhealthy for settlement. Having heard stories hailing beautiful country to the west, Griffith, accompanied by a few men, set out on horseback and eventually reached the Mad River. Following the water's swift-moving current they came upon James Demint's cabin nestled atop a hillside overlooking the valley below.

Kentucky surveyor John Dougherty was, by happenstance, already a guest at the cabin during Foos' arrival, but Demint nonetheless welcomed his family inside, where he told the new arrivals of his plans for a town south of the stream, just below the cabin. He then offered them a tract of land at a very low price that was duly accepted.

The men from Foos' exploration party returned to Franklinton, a journey of more than 45 miles, to gather their families and supplies, then returned to the west to begin life in a beautiful new valley. Griffith Foos built a double-log cabin in an area that today rests south of Main and just west of Spring streets. His home became the first within Demint's original town plat. Foos' log cabin and tavern opened in June of 1801, with friends and neighbors traveling from miles around to provide assistance and celebrate.

Operation of the tavern continued until May, 1814, when Griffith retired to his farm east of town. By 1817 he'd built and begun operating a small oil mill on the corner of Linden and Monroe streets, and remained active in public affairs, serving on township trustee committees from 1812-1830.

Griffith's wife, Elizabeth, died on December 10, 1833 at the age of 56 and was buried in Columbia Street's Demint Cemetery (Section I, Lot 7). Griffith married Nancy Winans, a widow, on October 23, 1834. She died on February 11, 1858 at the age of 60. Griffith himself soon followed. According to the *Mad River Valley News and Clark County Journal* dated May 11, 1859, "Griffith Foos, one of the earliest settlers of Clark County, died on Thursday, May 5th (in) Cincinnati." He was buried next to Nancy in Greenmount Cemetery. In the will of Griffith Foos, probated June 3, 1859 in Clark County, Ohio, much consideration was given to his nephews, Nelson Foos and heirs of Columbus, Ohio, Gustavus Foos and heirs of Springfield, Ohio, and eight children of Griffith Foos, deceased, late of Wilmington, Ohio.

Gustavus Foos, son of General Joseph Foos and nephew of Griffith, spent his early years on a farm before receiving his formal education in the Springfield Schools system. He eventually traveled to Illinois, where his brother, William, established a mercantile business. He was later employed

Gustavus Foos

john Foos

by his brother to work in a general store. With Brothers John and William, he established in 1859 the Private Banking House of Foos Brothers, which later became Second National Bank. By 1873, Gustavus began to manufacture wringers, and from this humble beginning grew the agricultural implement business known as Foos Manufacturing Company. He also headed the Cemetery Association committee that purchased the land known today as Ferncliff Cemetery, devoting both time and money to help beautify this new "city of the dead." He helped organize the Board of Trade and, during the Civil War, served on the county executive committee that organized the 44th Ohio Regiment.

Entrepreneurial brothers John, Gustavus and William were closely connected in most of their enterprises—from the small merchandising business that grew into Mast-Foos Manufacturing to the banking industry. William served as Second National Bank president from its inception, while John presided over Springfield National Bank until illness and age overcame him.

From the onset of Griffith's arrival in 1801,

the Foos men proved to be true pioneers, lifting Springfield to the manufacturing forefront by 1900. The family is buried in Ferncliff Cemetery, Section A, Lots 1-8. Griffith and Nancy Foos' bodies were transferred from Greenmount to Ferncliff in 1937.

David Lowry
(1767 - September 9, 1859)

David Lowry was born in Pennsylvania in 1767, the son of David and Lettice Lowry, both natives of Scotland. According to *Ohio Daughters of the American Revolutionary Soldiers, Volume 1,* he was a member of the provisional train that supplied General Anthony Wayne's army. He traveled to Ohio after his service and settled in Bethel Township (Clark County), Ohio in 1799, content to farm and build and operate grist mills. Lowry and Jonathon Donnel were likely the first families of settlers in the township.

In *Sketches of Springfield*, authored in 1852 by Robert C. Woodward, Lowry assisted two men in 1796 with "the names of Krebs and Brown" to raise "the first crop of corn in the neighborhood of Springfield." By 1800 he'd built the first flatboat "that ever navigated the Big Miami from Dayton down." It was aboard this boat that Lowry made the first-ever shipment of provisions to New Orleans. He also assisted in surveying the first public road to Dayton, as early as 1801 or thereabout.

Lowry married Sarah Hammer in nearby Miami County, Ohio in November of 1791 and the couple raised four daughters: Sarah, Nancy, Elizabeth and Susan, the last of which married John Leffel. Mrs. Lowry died in August of 1810. David remarried on February 14, 1811 to Jane Wright Hodge, the widow of Andrew Hodge. To the couple were born Martha S., David W., Robert M. and Sarah R. Lowry continued mill operation and farming until quite late in life. He died on September 9, 1859 at the age of 92 years, 10 months.

He was buried in Minnich Cemetery but later transferred to Ferncliff Cemetery, where he rests in Section D, Lot 12. Jane Lowry died on August 15, 1867 and is interred beside her husband.

George W. Clemans
(March, 1852 - February, 1861)

Unbeknownst to many visitors, one of Ferncliff Cemetery's most unremarkable stones holds a most remarkable story—that of eight-year-old George Clemans.

For more than a century, the whereabouts of his small grave remained a mystery. As time passed, the questions surrounding his death—and his historical significance as Ferncliff's first burial—only grew. Where was he? Had he slipped into anonymity, laid to rest without a marker? Or was it all, in fact, simply legend?

On a hot, humid afternoon in August of 2005, the answers came.

Ferncliff office assistant Scott Brown, accompanied by Paul W. Schanher, III, decided on a whim to search the grounds in Section A—the cemetery's oldest area and site of its earliest burials. To adequately prepare for an upcoming fall walking tour of the cemetery, one ultimate question had to be answered. Was George Clemans buried here?

Plot maps and burial records indicated that three gravestones had once been placed in a sloped area overlooking the road in Section A's southeast side. Yet only two were visible. Something must be awry, they thought. Undeterred and just hours prior to the cemetery's first tour, Brown and Schanher eventually noticed a subtle depression in the ground directly adjacent to one of two Clemans family markers. They quickly caught each other's eye with the same thought in mind: *What would cause that?*

Being the younger and more agile of the pair, Brown bent over to investigate. Locating a fallen branch from a nearby tree, he began to scratch away at the depression's surface. After digging some four inches below the surface, a white stone slowly emerged.

Their eyes widened. *Could this be the place?*

Eager with anticipation, Brown completely uncovered the hidden stone, and in the emotional process, restored identity and dignity to little George Clemans—buried first, and buried alone, nearly 150 years ago.

Today, his weathered grave, newly exposed to the light, reads, "George W. Son of John B. and C. Clemans... Died Feb., 1861." C. Clemans is most likely Clarrissa Clemans, whose tombstone remains upright and legible and sits between little George's and that of Calvin Clemans, who "died Dec. 21, 1873." Calvin's relation to the family remains a mystery, but 1870 census reports reveal that someone by that name lived in Springfield's Ward 5. All that is known of Clarrissa is preserved on her tombstone: "... wife of John B. Clemans. Born March 22, 1812, died Aug. 9, 1873."

The hearty souls who later toured the gently rolling hills of Ferncliff Cemetery unknowingly found themselves in the midst of history unveiled, being among the first to view the long, lost monument of George Clemans, who died of spotted fever at the age of eight years, 11 months, and 24 days. Interestingly, those numbers place his likely birth date in March of 1852. Clarrissa would have been roughly 40 years old at the time.

Today, the little boy is anonymous no more. His memory and his modest monument have been unearthed for all to see, marking the exact site of Ferncliff's first burial.

James Leffel
Mary Ann Croft
(April 19, 1806 - June 11, 1866)
(November 7, 1813 - October 27, 1887)

James Leffel was born April 19, 1806 in Botetourt County, Virginia and arrived in Clark County, Ohio with his family later that year.

Little is known of his early years, but as Leffel matured his interests gravitated to all things mechanical and he was eventually recognized as a natural mechanic and innovative genius, establishing Clark County's metal industry. According to Robert C. Woodward's *Sketches of Springfield*, Leffel established the county's first foundry west of Springfield on the south side of the National Road near Buck Creek bridge on January 1, 1840. The company manufactured axes, sickles and knives and performed iron implement repairs. Due to demand a second foundry was built on Buck Creek in June, 1845.

Leffel continued his inventions, producing small iron goods for the home and farm, most notably a lever jack, patented in 1850, and a variety of cooking stoves such as the "double oven" and "the Buckeye," patented in 1849. Two water wheels were patented in 1845 and 1862, including the double turbine, and by 1864 sold throughout America and abroad. Today, James Leffel & Company still bears his name and enterprising spirit.

He continued his mechanical enterprises until his death on June 11, 1866 at the age of 60. Mary Ann Leffel, the daughter of George and Mary Critz Croft, outlived the man she married on July 4, 1830 by more than two decades, dying on October 27, 1887. The Leffels and members of their family are interred in Section E, Lot 14 of Ferncliff Cemetery.

Daniel Hertzler
Catherine Hershey
(January 22, 1800 - October 10, 1867)
(January 12, 1809 - January 15, 1872)

Daniel Hertzler traveled to Clark County in 1834 from his Lancaster, Pennsylvania home, searching for greater opportunity to support his wife and infant daughter, Barbara. He settled at Snyder's Station and immediately purchased what was then known as the Menard Mill property nestled along the Mad River. Menard originally purchased the mill from James Leffel and added a brick distillery that proved profitable for several years. Unfortunately, he fell upon unfortunate circumstances and the property eventually landed in Hertzler's hands. He added a sawmill to its grist mill and distillery and called his operation the Hertzler Mills—a business that experienced prosperity for some 20 years.

Hertzler and his wife, Catherine Hershey, had 10 children, although two sons and two daughters died young. Those who remained eventually married into the Snyder, Huffman, Rubsam, Pope and Baker families. Hertzler decided in 1855 to sell his profitable property to Henry Snyder, move to Springfield, and enter the banking business. He established the Old Clark County Bank, reportedly the first private bank in the county, and while serving as principal owner and general manager, became well-known and well-connected in the moneyed world.

But a disputed investment eventually coaxed Hertzler out of the banking business. Much to his surprise, he returned one day from a business trip to find that a banking partner had invested $30,000 in railroad stock. Angered that as the company's primary stockholder he hadn't been informed, Hertzler promptly closed the bank and paid off the partner with ... the railroad stock he'd so unwisely purchased.

Perhaps disillusioned beyond repair, Hertzler returned to farming and mill operations on lands he'd purchased in Bethel Township. It was here that he also constructed a stately, bank-style house for his family—a most recognizable home today known as "Hertzler House" and located just inside the entrance to George Rogers Clark Park.

Eerie legend surrounds the historic home, for in the wee hours of October 10, 1867, a tragedy took place inside its walls. Awakened by the sound of intruders, Hertzler waged a ferocious fight with a gang of robbers and was felled by a gun shot. Probate records dated November 25, 1867 revealed that his newly widowed wife and grief-stricken heirs immediately authorized the offering of reward money in return for the arrest of those involved in the murder of Daniel Hertzler. Four persons were subsequently arrested and charged with the crime. Two of the suspects were acquitted and two escaped from an open jail door, never to be recaptured.

Hertzler was buried on the Old Snyder Homestead alongside a few of his children. Catherine died five years later in 1872. The Hertzlers were later re-interred in Ferncliff Cemetery, Section L, Lot 1.

General Samson Mason
Minerva Needham
(July 24, 1793 - February 1, 1869)
(December 5, 1805 - April 29, 1890)

General Samson Mason was born at Fort Ann (Washington County), New York on July 24, 1793. He fought in the War of 1812 and, at the age of 19, fought in the battle of Sackett's Harbor, where his colonel was killed. Upon discharge from the service Mason studied law in Onondago County, New York.

Mason came to Ohio in 1818. He traveled first to Cleveland, then Steubenville, and later, to Zanesville. He eventually settled in Chillicothe, where he was admitted to practice law in the state, and eventually arrived in Springfield from Columbus, making both his home and his professional living in Clark County.

Samson wed Minerva Needham, youngest daughter of well-known Dr. William Needham, on November 24, 1823. It was also during this time that he was elected to the lower house of the Ohio Legislature, where he served several terms and later moved to the senate. He was actively interested in the state militia and held numerous positions within the state service, including captain of a cavalry company. Afterward he worked his way to colonel, brigadier general, and major general.

During the Millard Fillmore administration, Mason served as United States district attorney for Ohio. He remained active until shortly before his death on February 1, 1869 at the age of 75. Newly widowed, Minerva Mason lived with a son, Colonel Edwin Cooley Mason of the regular army, stationed at Fort Ripley, Minnesota. She died at Fort Snelling, Minnesota on April 29, 1890 at the age of 84. A second son, Colonel Rodney Mason, was stationed at Washington.

Minerva Mason was removed to Springfield for burial beside her husband and other family members in Ferncliff's Section C, Lot 26.

Henry Snyder
Rachel Line
(December 19, 1783 - April 25, 1869)
(April 22, 1787 - June 26, 1865)

Henry and Rachel Line Snyder Sr. loaded up their children (John, David, William, Henry, Christian and Mary) and left their home in Cumberland County, Pennsylvania, moving westward in search of new opportunities. They settled near Dayton in 1824 and, within a year, Henry purchased from Elijah Harnett several hundred acres of Clark County land and a flourishing mill. The Snyder boys conducted the family's milling business and a second mill was erected and used primarily as a distillery. It was in operation until a heavy war tax in 1862 made business unprofitable.

Foreseeing the inevitable tax that would be placed on liquor, the brothers distilled a vast quantity before the tax was levied, making a huge profit from their timeliness and ingenuity. John and David L. eventually took control of the operation, with David serving as senior partner, and renamed the enterprise The D.L. Snyder Company. David was also associated with Victor Rubber Company, but the greatest of the brothers' holdings became real estate as the pair accumulated more and more land within the Mad River Valley.

As business prospered Henry Snyder Sr. purchased an additional 217 acres of land along Buck Creek in the western part of the county. Today known as Snyder Park, this land was donated in 1895 to the City of Springfield by sons John and David L. Snyder. Following John's death on December 14, 1896, and prior to David's passing on October 21, 1898, a $200,000 endowment in government bonds was also contributed for park maintenance. An additional $25,000 was later bequeathed for construction of a memorial bridge inside the park.

Henry Sr. died in his residence on April 25, 1869 at the age of 86. His wife, Rachel, preceded him in death in 1865. The Snyder children maintained their benevolence to their adopted City of Springfield in numerous ways, but are most remembered for the beautiful park they left behind. The Snyders are interred in a large family plot in Section E of Ferncliff Cemetery.

John Snyder

David Snyder

Captain William Addison Stewart
(May 25, 1809 - July 21, 1869)

William A. Stewart was born on May 25, 1809 in New York. He pursued river navigation at an early age, and by 1860, was making a living in Hamilton County, Ohio—first as as a river pilot, then as a captain on runs between Cincinnati, St. Louis and New Orleans. He was also a secretary and trustee of the Cincinnati and New Orleans Pilots' Association.

Stewart served in the Black Hawk War, holding command of the Illinois Volunteers. Prior to the Civil War in 1861, he was appointed by the government to an expert panel assembled to study the channels and bearings of the Mississippi River. Captain Stewart submitted his report in Washington, D.C. in July of that year. Upon his return and under the command of Commodore A.H. Foote, Stewart fitted out gunboats *Carondelet* and *Mound City*, taking command of the latter during several military maneuvers on the Mississippi River.

Under the direction of Admiral Porter, he was later engaged as a U.S.S. Osage river monitor pilot in the battle at Fort Durussy on the Red River. He tendered his resignation in April of 1864, having received a commission from the Secretary of the Treasury as a U.S. Local Inspector of Steamboats for the Cincinnati district, a title he held until his resignation in 1867.

Captain Stewart moved with his family to Springfield in July of 1866 and entered the coal business as a senior member of W.A. Stewart & Company. After the notorious murder of prominent Clark Countian Daniel Hertzler, Stewart was urged into the role of police chief on October 14, 1867. He later resigned the office and refused to accept payment for his service. In January of 1868 he became a stockholder in the Republic Printing Company.

Stewart died of heart disease on July 21, 1869 and was buried with his wife, Elizabeth, and other family members in Section D of Ferncliff Cemetery. Elizabeth died on August 26, 1860. The couple's son, Colonel James E. Stewart, died in 1889.

Colonel William Werden
Rachel Reed
(November 11, 1785 - December 17, 1869)
(1784 - July 25, 1860)

William Werden was born on November 11, 1785 in New Jersey, where he lived his early life before heading to Philadelphia to work in the wholesale leather business. He married Rachel Reed, the daughter of Robert and Sarah Green Reed of Trenton, New Jersey. The family traveled west in 1819 and settled in Clark County, Ohio. Eight children were born to the Werdens: Robert, Reed, James Duncan, William, Wharton, Sarah Ann, Mary Jane, and Rachel.

Werden launched his business ventures by opening Billy Werden's Tavern. After 10 years of profitable sales he opened the National Hotel, later renamed The Werden House. The establishment became a routine resting place for folks traveling to and from Columbus and Cincinnati by stage coach, and was utilized by prominent politicians of the day to promote their agendas. As a result, Werden became well-acquainted with Henry Clay, Tom Corwin and others. General Andrew Jackson, upon being elected the seventh president of the United States, appointed Colonel William Werden as Springfield postmaster.

Werden was involved with several others in establishing and operating the stage line between Springfield and Wheeling, and in 1860, worked as a banker. He was one of the founding members in December, 1834 of All Souls Parish in Springfield's Protestant Episcopal Church, and was a longtime Masonic member of Clark Lodge, #101 F. & A.M.

After the death of his wife, Rachel, in 1860, Colonel Werden lived with his son, William, in St. Louis, Missouri until his death from heart disease on December 17, 1869.

His obituary in the *Springfield Republic*, dated December 18, 1869, reported that Werden's body would be moved to the residence of his son-in-law, John W. Baldwin, on South Market Street for a funeral service on "Monday next at 2 o'clock. The deceased was a thorough gentleman of the old school, with strong points of character, and died at a ripe age, maintaining strength and vigor until a short time before his death. Few men were personally so popular, and a large circle of friends will mourn his loss."

Colonel Werden is buried alongside his wife in Ferncliff Cemetery, Section E, Lot 5.

Jonathon (Hug) Hook
Magdalena Brunner
(October 21, 1826 - February 14, 1871)
(December 29, 1830 - March 14, 1914)

Jonathon Hook was born in Milheim (Baden), Germany on the Rhine River near the French and Swiss borders. Nothing is known of his parents (nor any other family members), and it's uncertain when he arrived in America. An 1852 directory provides a clue, listing a John Hoak as a laborer living on Main Street between Race and Plum streets.

John wed Magdalena Brunner, of Baden (Baden), Germany, on September 23, 1856. Reverend Charles Stroud, M.G. performed the ceremony inside St. John Lutheran Church.

Ironically, nothing is known of Magdalena's background, but surviving ship passenger lists reveal that she arrived in New York Port as a 22-year-old on June 18, 1853 aboard the ship Robert, traveling from Havre. An 1860 Clark County census reports John Hook living in Ward 1 with a wife and daughter, age 2.

Once fear of a Civil War was realized, Hook enlisted in the 44th Ohio Volunteer Infantry being assembled at the Springfield fairgrounds during the summer and fall of 1861. On October 14, 1861, as well-organized troops marched through village streets en route to the front lines, huge crowds gathered to bid them farewell. John was forced to leave behind his wife and young children, Margaret, born December 13, 1858, and Frederick, born July 3, 1861.

The 44th was known as a single, continued advance and retreat, with constant skirmishing. By October of 1862 the troops were retreating from Lexington to Richmond, Kentucky when John fell and injured his foot and ankle. It was initially diagnosed as a strain until swelling revealed protruding bones. He was sent to General Hospital in Lexington and later transferred to Cincinnati's General Hospital, where he remained from February through April of 1863. He was discharged at Camp Dennison, Ohio on June 27th of that year, carrying in hand a Surgeon's Certificate of tarsal bone dislocation and chronic lung disease.

John filed an Invalid Pension Claim on February 8, 1864 and appeared on the Invalid Pension Rolls, with testimony provided by Captain A. Dotze, Michael Follrath and Nicholas Kalt. He signed the document "Johann Hug." Beneath his signature were the words, "English spelling John Hook."

Following his discharge, John Hook and his wife had two more children, John and Henry. As of 1870 the family lived in Ward 2, with John working as a laborer and his two oldest children attending school. The Hooks were active members of St. John Lutheran Church, which baptized all of the couple's children.

John died on February 14, 1871. He is honorably interred in Ferncliff Cemetery's Civil War Mound.

In 1869 the Grand Army of the Republic initiated plans to assist war veterans and, by 1871, the Ohio Soldiers' and Sailors' Orphans Home was opened in Greene County, Ohio. Hook's oldest son, Frederick, was admitted to the facility on March 11, 1872 (at the age of 11) and discharged in 1876. Records show that youngest son Henry was also admitted from September 3, 1873 to November 20, 1880, until a postal sent by his mother earned his release. No records have been found to indicate that John or Margaret were residents of the home, though they could have been.

Magdalena appeared in Clark County Court on April 11, 1879 to file a Widow's Claim for a pension. Appearing on her behalf were several friends and neighbors, namely, Jacob Kaiser, Nicholas Kalt, Jacob Leitschuh, Augustus Dotze (a late lieutenant in the 44th O.V.I. and a colonel in the 8th Ohio Volunteer Cavalry). Kalt repeated his testimony in 1880 and Mary Cool and Margaret Eyeper supplied statements. Curiously, records do not indicate when Magdalena began receiving her $12 monthly pension. By 1883 she was living on Cedar Street with sons John and Henry, who became a coremaker and a blacksmith, respectively. Neither remarried, living out their lives at that residence. John died in 1902, Henry in 1904, and Magdalena in 1914. Margaret married Cyrus Tabler and the couple had five children. Cyrus died in 1906 and Margaret in 1921. All are interred in Ferncliff Cemetery, Section L, Lot 121.

The only son to marry and extend the family name, Frederick wed Mary Biggins, daughter of Owen and Catherine McGivney Biggins, both natives of Cavan County, Ireland, in January of 1881. They had 13 children, 10 of whom reached adulthood. Mary died in July of 1927 and Frederick in January of 1932 following 44 years of marriage. Mary is buried among family in St. Bernard Cemetery. Frederick is interred in Ferncliff Cemetery, Section L. The last to be buried in this spot was Frederick Eugene Hook, great-great grandson of John. Born in 1947, he served in the Navy during the Vietnam War, returned to Springfield and worked as a Springfield firefighter, just as his father, Eugene, did before him. Frederick Eugene died in February of 2000.

Daniel Raffensperger
Mary Bowman
(1796 - October 20, 1877)
(1797 - May 28, 1872)

Among Clark County's most remembered families is the Raffenspergers, who traveled here from Pennsylvania. Daniel, the youngest of seven children, was born in York County and lived on the family farm until he reached maturity. He married Mary Bowman in 1822 and for the next 13 years raised a family and worked on a farm in East Berlin (Adams County).

The family arrived in Ohio in 1835 and settled in the western part of Clark County in the village of North Hampton, where Daniel operated a general store and established the first post office. He later initiated a second post office on Urbana Road called Prairie-Fountain at the Halfway House, where Daniel was the proprietor. When the Raffenspergers moved to town Daniel operated the Western House on Main Street, a popular stop for travelers heading west. He was elected sheriff in 1846, and by 1852 worked in the retail grocery business, according to old city directories. The Raffenspergers were active members of English Lutheran Church. Mary served as one of the founders of the organization in Springfield, and also as one of its "Mothers of Israel." There were at least five Raffensperger children: Edwin, Alfred, Henry J., John B. and Elizabeth.

Daniel Raffensperger died on October 20, 1877 in his 81st year. His beloved companion, Mary, preceded him in death on May 28, 1872 at age 75. They are buried together in Ferncliff Cemetery's Section D, Lot 10.

Reverend Edwin B. Raffensperger, the couple's oldest child, was born in East Berlin, Pennsylvania and spent his early years at home. Educated at Hanover and Princeton, he entered the Presbyterian Church ministry and was pastor of Presbyterian Church in Muncy (Lycoming

County), Pennsylvania at the time of his death on May 1, 1885. His body was returned to Springfield for burial.

From an early age, Alfred Raffensperger elicited an inventive talent and, at the age of 10, began manufacturing wooden matches that were sold at his father's general store in North Hampton. As a young man he worked in harness and saddle-making, joining his father in a saddlery business then located in Springfield's Driscoll Building. Alfred married Elizabeth Thomas on January 31, 1856 and the couple had several children: Clara, Mary, Lorene, Daniel and Edward. Daniel became well-known in 1889 when he established the Springfield Tea Company on East Madison Avenue.

During the 1860s Alfred entered real estate and platted the eastern part of Springfield. The district included Ludlow Avenue to Burnett Road, with one street in the addition bearing the name "Raffensperger." To promote sales he took out the first full-page advertisement ever seen during the spring of 1869. On the day of the sale, coaches were driven throughout the day, providing rides from town to the new development. The occasion was marked by musical entertainment as well, turning his real estate venture into a huge success.

Alfred died on October 6, 1916 at the age of 88. He is buried next to his wife.

Elizabeth was born in 1825 and married George W. Lipscomb on August 13, 1844. The pair had three known children. She died in 1878 and is buried in the family plot.

Little is known of Henry J. Raffensperger after he reached adulthood. He moved to Toledo, Ohio for a time working in photography. His death occurred suddenly from apoplexy September 26, 1888 at the age of 56 years in Springfield.

John B., the youngest of the family, was born in North Hampton, Ohio in 1838. With Civil War smoke hanging in the air, John became one of the first hundred men to respond to the President's call for 75,000 soldiers. He served for three months in the 2nd Ohio, commanded by Colonel McCook in the Company of Captain King. He was also in Company B, 86th Ohio Volunteer Infantry under Captain H.B. John.

Afterward, John enlisted in the 4th Ohio Volunteer Cavalry until war's end, at which time he was honorably discharged. He resumed his career as an ornamental sign and fancy cards painter until dying of consumption on February 2, 1882 at age 44. He is interred with the Raffensperger family in Ferncliff Cemetery.

Reuben Miller
Mary Hedges
(January 19, 1797 -October 3, 1879)
(1803 - January 2, 1875)

This pioneer family came to Champaign and Clark counties in 1812 when Reuben was just 15 years old. He was born to Reverend Robert and Mary Highfield Miller on the 19th of January, 1797 near Brownsville, Pennsylvania, where his parents settled during the winter of 1796-1797 while traveling from Virginia to Fleming County, Kentucky. The Millers later moved to a farm that, by today's boundaries, would be located within Clark County's Moorefield Township.

Reuben worked on the farm until the age of 22, after which he turned his attention to obtaining an education. Self-taught at times due to a shortage of teachers, he nonetheless gained considerably more than a basic education. A Miller biography found in Beers' *History of Clark County, Ohio, 1881* reveals that Miller "became a very good arithmetician, studied geometry, trigonometry, surveying (and) navigation, and acquired some knowledge of astronomy." He also "commenced the study of the Latin language, but failed for want of an instructor."

Reuben Miller married Mary Hedges, daughter of Samuel from Berkley County, Virginia, on March 27, 1823 and settled on a farm not far from his family. He taught school, farmed his property and conducted occasional land surveys until 1828, when he and Mary moved to Springfield. There, Miller was appointed to the County Clerk's office before being tapped by the Court of Common Pleas as Clark County surveyor, a position he held for nine years. He served as county auditor, one-term mayor and occasional justice of the peace during a 30-year period, juggling more than one position at various times.

Religion played a key role in the lives of the Miller family, passed on from his Methodist-minister father, who donated land to build the Moorefield M.E. Church. When Rev. Miller died on October 18, 1834, he was buried in the church's small graveyard. After the family moved to the city, Reuben Miller became one of the first organizers and members of the High Street M.E. Church. He was ordained a deacon in 1835, and as a local elder, preached twice every Sabbath for a number of years, holding that lifetime honor. Reuben and Mary Miller were married more than 50 years when Mary became ill and died on January 2, 1875. At the time of her death, seven of the Miller children were still living: D.B. Miller, M.D. moved to Kentucky, Mrs. R.B. (Elizabeth N.) Ogden and her brother Henry R. Miller relocated to Iowa. John C. Miller served as a Navy commander and Clark County probate judge. Joseph N. Miller was promoted to captain in 1881. A son, Robert Tabb, and a daughter, Caroline H., died earlier.

Reuben eventually moved to Keokuk, Iowa to be with his daughter, Elizabeth. He died there on October 3, 1879. His body was returned to Springfield and buried next to his wife and other family members in Ferncliff Cemetery's Section E, Lot 43.

Dr. Robert S. Rodgers
(September 11, 1807 - February 24, 1880)

Born in Cumberland County, Pennsylvania on September 11, 1807, Robert Rodgers was given early schooling and eventually graduated from the University of Pennsylvania in 1828. After practicing medicine for three years, he moved to Portsmouth, Ohio in 1832 and to Springfield the following year. Rodgers' brothers, William and Richard, and two half-brothers, Reverend James L. and Andrew Denny, also ventured to Clark County.

Rodgers made the practice of medicine his life's work. He was involved in the first Clark County Medical Society in 1838 as secretary and censor, but with few meetings held, the organization slowly went out of existence. On May 31, 1850 the present-day Clark County Medical Society was formed, with Dr. Rodgers twice serving as president. According to the *History of the Medical Society of Clark County, Ohio, 1815-1955*, authored by D.J. Parsons, M.D., Rodgers "had the reputation of being a very skilled surgeon. He read a paper before the Society giving an account of an operation he performed, being the first Caesarian Section ever performed in the County."

Dr. Rodgers and Effie Harrison, daughter of General John Harrison of Dauphin County, Pennsylvania, were married shortly before traveling to Ohio. Sarah Harrison, Effie's sister, married William Rodgers, brother of Dr. Robert Rodgers. Their children, John H., Richard H., Isaac W., Sarah H., and twins Frances and James, all led exemplary lives.

John H., the oldest child, pursued the medical profession after graduating from Miami University and the University of Pennsylvania. He entered practice with his father until the onset of the Civil War, when he served as a surgeon in the 44th and 104th O.V.I.—a period of more than three years. John returned to Springfield and resumed medicine until shortly before his death in 1908.

Dr. Robert and Effie Rodgers were staunch supporters of the First Presbyterian Church from 1834 until the organization of Second Presbyterian Church in 1861. Both remained in communion with the church until their deaths.

Dr. Robert Rodgers died from a lingering illness of several years on February 24, 1880. Effie followed on June 12, 1887. Family members are interred in Ferncliff's Section A, Lots 77-83.

John W. Baldwin
Rachel Werden
(December 25, 1807 - January 5, 1881)
(Dec. 28, 1819 - Feb. 7, 1903)

Among the first settlers to arrive in Champaign County, Ohio were the Baldwins in 1807. John W., one of eight children of Joseph and Elizabeth Wilson Baldwin, was an infant during their Ohio arrival, having been born in Garrettstown (Berkeley County), Virginia on December 25, 1807.

John's father was also born in Berkeley County, on July 7, 1773. He and Elizabeth are buried in Buck Creek Cemetery in nearby Champaign County.

As a young man, John Baldwin and his brothers entered the dry goods business, maintaining large stores in New York City in addition to those in Springfield and Columbus. They worked in the firm of Baldwin, Dibley and Work until John permanently established himself in Springfield as a partner in Baldwin & Company. He was also affiliated with the Mad River National

Bank, serving as president until his death.

John W. married Rachel Werden, one of eight children born to Colonel William and Nancy Reed Werden.

At an early age, Colonel Werden traveled with his family from Delaware to Springfield. He served as a Seminole War soldier at the age of 19, and soon after his Springfield arrival, erected and managed the National Hotel. Werden also helped establish and operate stage lines between Springfield and Wheeling and eventually served as postmaster. His son, Reed Werden, became an Admiral in the United States Navy.

John and Rachel Baldwin were blessed with six daughters: Elizabeth, Sarah, Clara, Mary, Laura and Eleanor. Their oldest, Elizabeth, married Samuel F. McGrew—his family being among Clark County's more well-established. Sarah Louise was married in December of 1873 to John A. Blount, a descendent of early pioneers. Clara wed Thomas F. McGrew, while Mary Hepza married H.H. Moores, son of William and Elizabeth Cobb Moores, a well-known and honored family, in March of 1886. Laura married Montrose B. Wright, son of Dr. M. Wright of Cincinnati. Eleanor wed Douglas Hollister of New York.

John W. Baldwin remained active in business affairs until his death on January 5, 1881 at age 73. Rachel lived the remainder of her life surrounded by her children and grandchildren until her death on February 7, 1903. She was 83 years old. They and eleven family members rest in peace inside the Ferncliff Cemetery's Baldwin Mausoleum, Section E, Lot 8.

E.C. Middleton
Mary J. Lovejoy
(July 2, 1818 - May 13, 1883)
(1824 - August 22, 1885)

Elijah C. Middleton was born in Philadelphia, Pennsylvania on July 2, 1818. He was schooled during his early years and, at the age of 12, was apprenticed to a copper and plate engraver, eventually mastering the trade. He came to Cincinnati, Ohio in 1847 and joined Middleton & Wallace, Lithographers as a senior member, later founding Middleton & Strobridge. Middleton originated chromo-lithography as applied to portraiture and gained fame for his work in 1860-1861 via the publication *Middleton's National Oil Portraits* for oil chromos of George and Martha Washington, Henry Clay, Daniel Webster, Ulysses S. Grant, Abraham Lincoln and others.

Shortly after he settled in Cincinnati he met

and married Mary J. Lovejoy. The couple had two known children, namely, Edward C. and Mary J. "Nettie" Middleton. Mary married attorney Charles E. Evans on November 4, 1869.

During the Civil War young Edward was anxious to join the Union Army. Once he reached the age of 15, Middleton received parental consent and, on January 15, 1864, enlisted as a bugler in Troop #1 of the 4th Ohio Cavalry. Once reaching the army camp, he was assigned as a private, as there were no openings for buglers. He informed his father, who immediately wrote to President Lincoln to request his son's discharge. The President replied that his request would be granted, so long as a substitute was provided. These letters remain to this day and may be viewed at the Clark County Historical Society, where they're preserved in the museum's military gallery.

Private Middleton was moved with General Sherman's army into Georgia and later captured during a Confederate raid on June 21, 1864. He was confined in Andersonville, a particularly infamous prison camp, where he remained until war's end.

The family moved to Springfield in 1866 and Mr. Middleton continued his lithographic work under the name E.C. Middleton & Company. Many of his advertisements can be found in the newspapers of his day. He also became involved in the real estate business and Champion Malleable Iron Works.

Edward C. continued living in the family household until his marriage to Mary C. Cavileer, daughter of early Clark County pioneers Alfred and Iserelda Gordon Cavileer. They eventually moved to Washington, D.C., where Edward died on December 4, 1908. Mary later returned to Springfield, where she died on January 14, 1944.

Elijah C. Middleton died at his residence on May 13, 1883. His wife passed away on August 22, 1885. Family members are interred in Ferncliff Cemetery, Section H, Lot 161.

John Ludlow
Elmira Getman
(December 8, 1810 - June 10, 1883)
(March 27, 1814 - June 17, 1901)

Among the earliest of Clark County pioneer families were the Ludlows, who traveled here from New Jersey, lived briefly in Hamilton County, and in 1804 settled west of Springfield with the Reeder family and established a tannery. Son John Ludlow was born to Cooper and Elizabeth Reeder Ludlow on December 8, 1810, one of five children. One child died in infancy and soon after, Mrs. Ludlow passed away as well. Cooper Ludlow married again on December 14, 1813 in Champaign County, this time to Elizabeth Layton, daughter of Joseph Layton. They had nine children.

John Ludlow received his schooling from the primitive schoolhouses of the time. As he matured he became interested in the pharmaceutical trade, and under the leadership of Moses M. Hinkle, later completed his education in the employ of Goodwin & Ashton of Cincinnati. He joined Dr. W.A. Needham in business once he returned to Springfield and, after Needham's death, was associated with Cyrus T. Ward before forming a business partnership with Joseph Wheldon. Ludlow eventually bought out Wheldon for sole ownership. Ludlow married Elmira Getman, daughter of Frederick and Mary Getman, of Herkimer County, New York, on August 31, 1835. The Ludlows had three children. Daughter Ellen later became Ohio's First Lady when her husband, Asa S. Bushnell, was elected governor in 1896. Son Frederick established himself in California business dealings, while sibling Charles succeeded his father in Springfield pharmaceutical operations.

Beloved and highly respected, John Ludlow was elected as a director of the Springfield Bank in 1851, and later became its president with the death of Oliver Clark.

Deeply concerned about local cemetery neglect and overcrowding, Ludlow worked tirelessly during the 1860s to rally public interest in creating a more peaceful and elegant place of rest that Springfielders could take pride in and visit more comfortably. A man of vision, Ludlow imagined serene, landscaped grounds in the tradition of the recent rural cemetery movement begun in Mount Auburn, Massachusetts.

Ferncliff Cemetery was thus dedicated on July 4, 1864, with Ludlow serving as one of its inaugural directors. Notably, he also served as president of its board of trustees far longer than anyone who followed him, thoroughly devoted to his vision. His creation was soon recognized by people here and abroad as one of America's most striking and pioneering places of rest.

Ludlow's keen interest in Clark County history was revealed when he presented a paper he'd authored entitled, *The Early Settlement of Springfield* during a Mad River Valley Pioneer and Historical Association meeting in 1871. The booklet remains a classic even by today's standards, labeled an excellent source of early history that spans a period of 70 years. The work is sometimes referred to as *The Ludlow Papers of 1871*.

John Ludlow died on Sunday, June 10, 1883 at the age of 72. The *Springfield Daily Republic* of June 11, 1883 wrote, "As he lived so he died; thoughtful in the behal of others; solicitous for the welfare and prosperity of his fellows." Elmira Ludlow died on June 17, 1901. They are buried together in Section B, Lot 9 of Ferncliff Cemetery.

Rev. Joseph Clokey, D.D
Jane Patterson
(December 25, 1801 - December 8, 1884)
(1809 - 1832)

Elizabeth Waddle
(1808 - 1889)

Family records dating to the 17th century reveal that the Clokey family came from Ireland, landed in America on March 3, 1799, and settled in Dauphin County, Pennsylvania. Joseph (1754-1826) and Elizabeth Mitchell were married prior to their arrival in America and raised seven children. His second marriage, to Mary Sawyer, widow of William Sawyer, in Pennsylvania produced five children, Mary, Joseph, Andrew (who died in infancy), John Sawyer and Eliza.

Joseph Clokey was born in Dauphin County

on December 25, 1801. He spent his early years helping with family chores and obtaining a liberal education, graduating from Jefferson College of Canonsburg, Pennsylvania (1823). He was licensed by the Presbytery of Chartiers on July 4, 1826 and first ministered at Peter's Creek Church in Washington County.

Clokey married his first wife, Jane Patterson, on October 3, 1827 and the couple had one son and one daughter. Son John Patterson Clokey died on January 9, 1856 at the age of 23 and is interred in Springfield's Greenmount Cemetery.

On February 21, 1838 he married his second wife, Elizabeth Waddle of Short Creek, Virginia, and the couple had four sons and two daughters. The family moved to nearby Xenia, in Greene County, Ohio, where Joseph continued farming and pursued his interest in theological studies. He became Professor of Pastoral Theology and Sacred Rhetoric in the Xenia Theological Seminary, and later received a call from St. Clair, Pennsylvania, where he preached for seven years.

Following the resignation of Reverend Robert Henry from United Presbyterian Church in Springfield, Ohio on June 9, 1853, the church's congregation found itself without a pastor until the fall of 1854, Dr. Joseph Clokey, brother-in-law of Reverend Henry, was called. At the time, he was considerably past middle age, yet he remained active, influential and fearless in promoting the Union and anti-slavery sentiment during the Civil War. Due to increased infirmity, Dr. Clokey was compelled to retire from his pastorate and resigned on March 1, 1875, having served for 20 years.

After retirement, he continued his work with the Bible Society of Springfield and Clark County, an auxiliary of the American Bible Society of which Dr. Clokey served as president from 1860 through 1863. He died of renal congestion on December 8, 1884. Elizabeth survived him until July 4, 1889.

Joseph Waddle Clokey, eldest son of Joseph and Elizabeth, was born February 22, 1839 in Jefferson County, Ohio. He was well educated, graduating from Jefferson College in Canonsburg, Pennsylvania (Class of 1862), and was ordained a United Presbyterian Minister on September 7, 1864. Clokey served as pastor in Richmond and New Albany, Indiana, as well as Middletown and Steubenville, Ohio. He married Dellie R. Ekin on October 6, 1864, and raised two daughters, Allison and Gertrude. Dellie became ill at their New Albany residence, returned to Springfield, and died in the home of Reverend Dr. Clokey on February 16, 1881 at the age of 39.

On December 30, 1885 Reverend Joseph Waddle Clokey married Florence Day of New Albany, Indiana. They had one son, Joseph Waddle, born August 28, 1890 in New Albany. As he grew to manhood he soon graduated from Miami University, becoming an educator, organist and composer of sacred and secular music. Today he is known as one of the most widely sung composers throughout American churches.

He served as dean of the School of Fine Arts at his alma mater, Miami University (1939-1946), and also once served as professor of organ at both Miami and Pomona College in Claremont, California. He lived a full life, dying on September

14, 1960 in Covina, California.

Rev. Joseph W. and Florence Day Clokey lived out their lives, with Joseph dying on August 17, 1919 in Oxford, Ohio and Florence on December 17, 1924 in New Albany. All are buried in Ferncliff Cemetery.

Alexander Wilson Clokey, second son of Rev. Joseph and Elizabeth, was born in Jefferson County, Ohio, graduated from Wittenberg College in 1864, then studied theology in Xenia. He served as pastor in Indianapolis, Indiana (1867-1868), Aledo, Illinois (1869-1872), Butler County, Ohio (1874) and at the Presbyterian Church in New Carlisle, Ohio. He and his wife, Frances Chase Clokey, raised four children.

Josiah Mitchell Clokey, third son of Rev. Joseph and Elizabeth, served in the Civil War (Company E, Bushnell's 152nd O.V.I.), and later became a lawyer, married and raised two known children. Two of his daughters married: Mary wed James Porter and Anna wed John Drury. There were no known children born to either woman.

The Clokey family is interred in Section I, Lot 9 in Ferncliff, with many members having been transfered here from other burial sites in the United States.

John Funk
Martha Kauffman
(1808 - 1888)
(1811 - 1860)

The John Funk family migrated to Ohio from Lancaster County, Pennsylvania in 1834 and settled in the small town of Clifton, Ohio. By 1840 the Funks moved to Springfield and settled in the Pennsylvania House, a newly constructed inn and tavern that served as an important rest stop for weary travelers heading west along the National Road. A liquor merchant by trade, Funk and his wife took over operation of the tavern for the next several years.

Early newspaper man George W. Hastings, publisher of the *Nonpareil* in 1853, related a story in 1912 about his 1845 stay with Christopher, John and Henry Funk at the Pennsylvania House upon his arrival from Cincinnati. He was given a nice, clean bed in an airy room and the next morning Mrs. Funk served a delicious breakfast with warm biscuits and butter.

By 1850 there were eight Funk children, five

boys and three girls. During the next 10 years, oldest son Henry J. and Christopher C. entered the merchandising trade, operating a wholesale grocery at 71 West Main Street in Springfield (later followed by Benjamin F. and John A., all at the same address). John A. also pursued the dry goods business with R. Stanley Perrine. The Funk & Perrine store was located at 67 W. Main Street. Elder daughter Barbara E. married Urbana native George Steinbarger and the couple had three children, all enterprising in their various activities.

Following the death of Martha, John Funk's first wife, he married again and fathered one child, Lovetta, who died at the age of 15. John passed away years later, on June 25, 1888, at the age of 80. He is buried among family members in Ferncliff Cemetery, Section H, Lot 118.

Isaac Kauffman Funk, fourth child of John and Martha, was born September 10, 1839 in Clifton, Ohio. In 1856 he entered Wittenberg College and Theological Seminary in Springfield, graduating in 1860 as a Lutheran minister. He pastored throughout Indiana, New York and Ohio, and by 1872 was touring Europe, Asia and Africa.

He founded I.K. Funk & Company publishing company in 1876 and, with college classmate Adam W. Wagnalls, an 1867 graduate, printed the *Homiletic Monthly*, a magazine serving ministers and religious workers. *The Literary Digest* was also published that same year.

The enterprising pair later published dictionaries and encyclopedias, with Funk's most notable achievement being the *Standard Dictionary of English Language*, published in 1893. From 1901 to 1906, Funk and Wagnalls compiled the *Jewish Encyclopedia*. Other publications included *The Next Step in Evolution* (1902), *The Widow's Mite and Other Psychic Phenomena* (1904), and *The Psychic Riddle* (1907).

Dr. Funk died April 4, 1912 in New York and is interred in Brooklyn's Green-Wood Cemetery, Section 166, Lot 26127.

Broadwell Chinn
(1855 - September 11, 1889)

Broadwell Chinn was born around 1855 in Lexington, Kentucky. His large family arrived in Springfield during the 1870s, where father William opened a Center Street grocery store. Broadwell attended primary school but racism later denied him admission to public high school. However, tutoring he received from a principal and several teachers allowed him to qualify for a high school diploma. He soon applied for admission to Wittenberg College in September of 1874. According to the *Daily Republic* newspaper, in an article dated September 11 of that year, Wittenberg's faculty took his application under advisement, opting to relegate the matter to a vote of the student body, which, sadly, rejected him. Wittenberg's Board of Directors, however, reversed that vote and admitted Chinn.

An 1877-1878 city directory lists Chinn's occupation as "law student." Later directories identify him as a musician and teacher. A man of obvious courage, intellect and tenacity, Chinn became the first African American to graduate from Springfield High School and Wittenberg University, and also broke racial barriers as a member of the Clark County Bar Association.

His cause of death listed as "consumption," Chinn died on September 11, 1889. He is buried in Section F, Lot 90 of Ferncliff Cemetery, without a stone to mark his grave. His father, William Chinn, died in 1891. A small marker identifies his place of rest.

Samuel A. Bowman, third from left

Samuel Andrew Bowman
(1832-1895)

Samuel Andrew Bowman was born in Zanesville, Ohio in 1832. He came to Springfield, by way of the National Road, to attend Wittenberg College, graduating in 1852. Following graduation he remained in Springfield and began reading law in the office of General Sampson Mason. Shortly after he began his career, he was married to Adaline Ogden (1833-1895), daughter of Edmund Ogden (1789-1868) and Adaline Moore Odgen (1808-1884) in 1857.

Bowman was admitted to the bar and, in the days preceding the Civil War, began practice with James Samuel Goode. At a later date he was in practice with Samuel Shellabarger, a prominent lawyer and eventual congressmen. Bowman served as corporate laywer for William Whitely, whose company produced the famed Champion Reaper and Mower. Near the end of his life he went into practice with three sons - Edmund O., John Elden, and Border.

Bowman was a civic-minded individual who often lent his experience and expertise to various community endeavors. He served as a member of the Board of Directors at Wittenberg College from 1861-1883. In 1864 Bowman served as chairman of a committee that raised an additional endowment for the college of one hundred thousand dollars. Bowman also founded the Springfield Seminary which served as one of Springfield's premier institutions for religious education.

Samuel Bowman was also instrumental in the creation of the Ferncliff Cemetery. He played a critical role in developing the legal framework that would help establish and maintain the operations of the cemetery.

He was also active in several fraternal organizations and was the founder of the Literary Club, which held the first meeting at his house located at 815 East High Street.

Bowman died in Springfield in 1895 and rests in Section C, Lot 9 in Ferncliff Cemetery.

Samuel Shellabarger
Elizabeth Brandiff
(December 10, 1817 - August 6, 1896)
(July 11, 1828 - December 3, 1914)

Samuel Shellabarger was born on December 10, 1817 in Mad River Township (Clark County), Ohio to Samuel and Bethany McCurdy Shellabarger, the oldest of eight children. His early life was spent on the family farm, completing the rough, daily tasks of early pioneer times. He grew weary of farm life as he matured, and, recognizing his intellectual interests, Shellabarger's parents encouraged him to study and forge a career.

He began his early training in the old Presbyterian church near Enon. Later, with his brothers and sisters, he attended some of the early schoolmasters' classes held in the area. With the opening of Springfield High School, he enrolled to prepare for college. He entered Miami University in Oxford, Ohio in 1837 and graduated four years later, then immersed himself in the study of law with Samuel Mason, a highly respected Springfielder. Samuel was known as a studious young man with a keen mind, qualities that helped admit him to the bar in 1846.

While practicing law in Troy, Ohio, he met and married Elizabeth Brandiff on May 25, 1848 and the couple had several children. That same year, he opened a law practice with James M. Hunt.

Shellabarger was politically motivated, elected to the General Assembly in 1852 and to the United States Congress in 1860, serving several terms. Upon his retirement from Congress, he moved to Washington, D.C. and continued his law profession, engaged in litigation with the U.S. Supreme Court. Ill health plagued him most of his adult years but didn't deter him from his elected duties and responsibilities.

Son, Robert, also an attorney, joined Shellabarger in his law practice until his death from yellow fever in 1889. After the subsequent death of his daughter-in-law in 1893, Samuel Shellabarger provided the care and upbringing for his grandson and namesake, Samuel.

Young Samuel received a formal education and became a teacher, serving as headmaster at Columbus School for Girls until 1946. He later retired to Princeton, New Jersey to devote his time to writing. His novels were read worldwide and transformed him into one of America's leading historical writers. His works included *Captain from Castile*, *Prince of Foxes*, *The Black Gale*, *The King's Cavalier*, *The Token*, *Lord Vanity*, and *Tolbecken*.

Judge Shellabarger, as he was affectionately and respectfully known, died in Washington, D.C. on August 6, 1896 following a lingering, months-long illness. His body was returned to Springfield and is interred with his wife and other family members in Ferncliff Cemetery's Section B.

Charles Albert Cregar
Mary Burns
(May 8, 1858 - July 8, 1896)
(November, 1859 - June 25, 1942)

Charles Cregar was born in Springfield, Ohio on May 8, 1858, the second son of Nathaniel and Catherine Smith Cregar, who hailed from Washington County, Pennsylvania.

After graduating from Springfield High School, Cregar studied draftsmanship, subsequently traveling to Fort Wayne, Indiana to study the trade under noted draftsman T.J. Tolan. Returning to Springfield, he was hired as the principal draftsman for T.B. Peet & Company, which manufactured galvanized iron fixtures.

Cregar married Mary Burns, daughter of Edward and Ellen Burns, natives of England and Ireland, on November 7, 1879. The couple had five children.

In 1884 Charles went into business with his father, a well-known contractor and builder, becoming a partner in an architectural business the pair opened in the Mitchell Building. There they designed some of Springfield's finest buildings, ironically, structures built "to last forever" but which, over a period of many years, were eventually torn down.

The list of buildings for which Charles Cregar served as architect is both lengthy and impressive, and, among others, includes the City Building, Saint John's Evangelical Lutheran Church, Clark County Courthouse, First United Presbyterian Church, Saint Joseph Roman Catholic Church, Saint Raphael Church and First United Presbyterian Church.

Upon the death of his father in 1885, Cregar and his brother, Edward, maintained the firm until Charles' sudden death on July 8, 1896 at the youthful age of 38. He is interred in the family plot, Section L, Lot 133 of Ferncliff Cemetery.

Charles Cregar was a member of Clark Lodge #101 F. and A.M., Springfield Council #17 R.A.M., Palestine Commandery #33 K.T., Springfield Lodge #33 I.O.O.F., and Moncrieffe Lodge #33 K.P.

Robert C. Woodward
Lizzie A. Crooker
(June 3, 1829 - July 24, 1896)
(August 8, 1832 - June 1, 1865)

Harriet DeWitt
(February 20, 1840 - December 29, 1912)

Jacob Schenck Woodward, a native of Chester County, Pennsylvania, came to Springfield, Ohio as a young man during the early years of the village's settlement. He pursued his trade as a spinner and weaver, and later joined Ira Paige and James Taylor in the operation of a woolen cloth mill.

On February 19, 1826 Jacob married Sarah Christie, the daughter of Major Robert and Sarah Goodrich Christie. The Christies had traveled from New Boston (Hillsboro County), New Hampshire to Springfield a few years before, in 1817. The couple had two children, John W. in January, 1827 and Robert W. in June, 1829.

Jacob partnered with John I. Wallis in the mercantile business, and they were in the process of expanding their company to West Liberty (Logan County), Ohio when Jacob died unexpectedly from bilious fever in September of 1829, just 29 years old.

Sarah assumed the solo task of rearing her two sons and, given her education as a teacher, earned money for their support and provided them with an excellent, rudimentary education. After her second marriage to John D. Nichols, a well-known publisher and journalist, in 1837, son Robert attended Ohio Conference High School and was subsequently one of the first students to enter Wittenberg College. He became employed in the newspaper business, working for the *Republic* and, later, for the Cincinnati *Commercial* as a compositor. At the same time, he pursued a commercial course with R.S. Bacon, then worked for two years as a bookkeeper in Davenport, Iowa. He returned to Springfield and went into business as a bookseller from 1859 to 1861, later pursuing the same occupation in Lima. In 1869 Robert and his step-brother, W.G. Nichols, repurchased his old book store and added a printing office before he retired from the firm.

The Springfield Library was housed in Union Hall until 1890, when it was moved to its present location. The Warder Free Library was thus dedicated June 12th with Robert Christie Woodward serving as librarian. He was the first known man to leave any detailed history of Springfield, authoring *Springfield Sketches* while compiling historical data for the inaugural edition of the *Springfield City Directory*, both published in 1852. He continued as librarian until his death on July 24, 1896, having been associated with Warder Library throughout its organized history.

Mr. Woodward married twice. His first marriage was to Elizabeth Abbie Crooker, a Chelmsford, Massachusetts native, on April 10, 1860. Lizzie taught in the Springfield schools and was actively involved with the Lima-based Women's Christian Commission during the war. After Lizzie's death in 1865, Woodward married Harriet DeWitt of Wyandot County, Ohio, on October 10, 1876. She taught school prior to their marriage and died on December 29, 1912. They are interred in Section C, Lot 21 of Ferncliff Cemetery.

Addison W. Butt
Frances G. Bagley
(October 14, 1835 - August 10, 1897)
(March 18, 1840 - March 12, 1927)

The Wendell Butt family emigrated from Germany and settled in Pennsylvania, where Wendell's son, George (and father of Addison) was born. George was a farmer by occupation and a master of the miller's trade. He married Olive Wyllis, with son Addison being born on October 14, 1835 in Chantanqua County, New York.

The family moved to LaPorte, Indiana in 1836, where George continued his agricultural pursuits until the death of his wife in 1838. He then pursued the milling trade until 1954, establishing a general store with son Addison under the name George Butt & Son. Addison established an agricultural implement house in LaPorte in 1861, and later became associated with the firm of Springfield-based Thomas & Mast, traveling as far east as New York and as far west as Nebraska until the firm dissolved. He was also known as an early member and stockholder of P.P. Mast & Company.

Addison Butt and Frances G. Bagley of LaPorte, Indiana were married on November 4, 1862 and had nine children. Frances was born March 18, 1840 in Mercer, Pennsylvania, the daughter of Asher F. and Elizabeth Bagley.

Butt was elected president of the Springfield Athletic Club in 1882, at which time the facility was greatly in debt. After three years under his management the club's debt was liquidated, with over $1,000 in the treasury. He also served as a charter member of Anthony Lodge #45, F. & A.M., Springfield Chapter #48, R.A.M., and Palestine Commandery #33, K.T.

Addison joined the Albion Manufacturing Company of Albion, Michigan in 1886, serving as president. In 1889 he purchased the old Springfield Manufacturing Company and forged it into a stock company, reorganizing under the name "Springfield Implement Company" while serving as its president.

Mr. Butt became ill in September of 1896 and was treated in both Springfield and Cincinnati for a disease diagnosed by physicians as "paralysis of muscles and nerves." His death occurred on August 10, 1897. Wife Frances outlived him by 30 years. She died at the home of her daughter, Mrs. James Turner, 2125 East High Street.

They and other family members rest in Ferncliff's Section K, Lot 1.

Phineas P. Mast
Anna M. Kirkpatrick
(January 3, 1825 - November 21, 1898)
(1826 - April 21, 1895)

Originally from Lancaster County, Pennsylvania, the Mast family moved westward to Ohio in 1830, settling in Champaign County.

Phineas was about five years old at the time, having been born on January 3, 1825. The Mast clan settled on a farm near Urbana where Phineas, accompanied by his four brothers and three sisters, tended to daily chores and attended school.

Mast graduated from Ohio Wesleyan University in 1849, specializing in courses devoted to scientific and Biblical studies. He then returned to the family farm, entered the grain and farm produce trade, and taught school. On January 3, 1850, his 25th birthday, he married Anna M. Kirkpatrick, and they remained on the farm for six years until they relocated to Springfield in 1856. There, he formed a partnership with John H. Thomas for the manufacture of agricultural implements. Mast bought out his partner in 1871 and established P.P. Mast & Company.

He founded a second firm, Mast-Foos Manufacturing Company, in 1876 that produced wind engines, pumps, plows and mowers, and later incorporated P.P. Mast Buggy Company. He soon envisioned a magazine that could promote his products, hiring John S. Crowell to launch the endeavor and nephew T.J. Kirkpatrick to provide editing. The result of their teamwork was *Farm and Fireside*, a 16-page agricultural journal initially published in October of 1877. The firm became known as Mast, Crowell & Kirkpatrick, and its innovative journal gained a circulation of more than 200,000 farmers throughout the Midwest. In 1883 the firm acquired *Woman's Home Companion*, already an established success.

Mast later served as Springfield National Bank president when the organization was founded in 1882. He was a 22-year member of city council, serving as that body's president for many years, and governed as mayor of Springfield from 1895 to 1897. He also devoted time as Board of Trade president. Mast was equally prominent in church activities, serving for nearly 40 years as Sunday school superintendent in Central M.E. and St. Paul M.E. churches, giving generously to support charitable causes.

Anna Mast died on April 21, 1895, and her husband followed on November 21, 1898. The two produced no children, but upon the death of Mast's brother, Isaac, they adopted their three nieces.

Phineas and Anna Mast are interred with family inside the Mast mausoleum, Section O of Ferncliff Cemetery.

Hezekiah R. Geiger, Ph.D, D.D.
Nancy Hartford
(January 10, 1820 - July 18, 1899)
(1831 -September 30, 1900)

Hezekiah's life began in Greencastle (Montgomery County), Pennsylvania on January

10, 1820, one of 12 children born to Henry and Julia Rheubush. His paternal grandfather, Charles Geiger, emigrated from Germany and settled in Montgomery County, Pennsylvania about 1722. While living in Philadelphia he served in the Revolutionary War, and afterward, resumed his trade as a miller.

Henry, the father of Dr. Hezekiah Geiger, was born in Montgomery County, Pennsylvania on May 7, 1789. During the War of 1812, he joined General Scott's division, participated in the battles of Chippewa and Lundy's Lane, and was with Commodore Perry during his naval victory on Lake Erie, which anchored at Put-in-Bay.

At the close of the war Henry settled in Franklin County, Pennsylvania, and in 1815 married Julia Rheubush, born and reared in Hagerstown, Maryland. The family moved to Columbiana County, Ohio in 1833, then to nearby Holmes County until 1851, when they relocated to Urbana. Julia died on August 31, 1854, and Henry spent the remainder of his life with his children until his death on April 7, 1862. They are interred with family in Ferncliff Cemetery, Section G. Of this family of 12, several grew to become prominent professionals: two were ministers of the gospel, one was a physician, three more entered law, two were elected judges and one served as a Union Army general.

At the age of 13, Dr. Geiger moved to Ohio with his parents and siblings. He continued his education there, but completed his collegiate coursework at Pennsylvania College at Gettysburg (today known as Gettysburg College) in 1846. After graduation, he returned to Springfield to work with Dr. Ezra Keller, D.D. and Michael Diehl and assist in the founding of Wittenberg College. Dr. Geiger accepted a professorship with the new institution, teaching Latin, mathematics and natural sciences. In 1873 he devoted his entire time to the natural sciences and traveled extensively on scientific investigations. He resigned his Wittenberg faculty position in 1882 to further pursue work in the natural sciences. Early in his career, Dr. Geiger was ordained to the ministry by the Wittenberg Synod.

On December 14, 1854 Dr. Geiger wed Nancy Melvina Hartford, a West Virginia native educated in Steubenville, Ohio. She was a teacher at Springfield's Presbyterian Seminary when she met Dr. Geiger, and the couple raised seven children, six of whom graduated from Wittenberg. A daughter, Mary Alice, became Wittenberg's first woman graduate, completing her coursework in 1879 and, in the process, forever opening the college's doors to successful coed instruction. Her memory is honored annually with presentation of the M. Alice Geiger Award—bestowed upon a senior female student for outstanding contributions to the Wittenberg campus.

Dr. Geiger died on July 18, 1899 and his wife, Nancy, on September 30, 1900. Some of the Geiger children continued to live in the family home on Ferncliff Place, built in 1853 by their father—an historic dwelling that also served as a hideaway for fugitive slaves escaping through Xenia during the movement to free southern slaves. According to Wittenberg University public relations information, "Geiger's wife, Nancy, provided refuge for slaves in the barn."

The Geiger family tombstone peeks from between two shrubs in Section L, Lot 5 of Ferncliff, not far from the Hertzler family plot. Nearby are the modest markers of Hezekiah, Nancy, and a few of their pioneering daughters.

James Fleming
(April 18, 1825 - November 10, 1899)

James Fleming was born in Westmoreland County, Pennsylvania on April 18, 1825 and moved with his family in 1833 to Ohio, where his father, unfortunately, died. James' mother, however, continued on and established the family in Springfield.

Fleming's farming skills supported the family until he was 20 years old, at which time he apprenticed himself to a local plasterer to learn a new trade. By 1853 he was elected constable, but ill health forced him to resign two years later. He traveled to the warmth of California to recuperate, returned to Springfield a year later, and labored as a plasterer until 1860, when he was elected county sheriff on the GOP ticket. Fleming served in that capacity for four years before being appointed to fill a vacancy in the office of mayor. He was elected to that capacity the following spring, but ill health prematurely ended his term.

After traveling to Minnesota, he eventually returned to Springfield in 1869 and as an Assistant United States Marshal, assisted in gathering the 1870 census. He was appointed police chief in 1871 at the age of 46, and under his vigorous and wise management that particular branch of government became remarkably efficient. He remained in this position for five years.

A man of all skills, Fleming worked for Superior Drill Company until 1878, after which he was appointed superintendent of the Clark County Infirmary. In that capacity, he received assistance from his wife, Sarah McIntire (1822-1894), whom he had wed in November of 1845. Fleming served as infirmary superintendent for several years before his death in Springfield on November 10, 1899. He was 74 years old and survived by a son.

Recorded as "Flemming" in burial card indexes and plot records, he and his family rest beneath tombstones bearing the name "Fleming" in Ferncliff Cemetery, Section L, Lot 98.

Charles Rabbitts
Margaret Robison
(September 2, 1820 - December 16, 1900)
(1826 - July 2, 1903)

The George Rabbitts family immigrated to America from England in 1832, seeking a better life. On January 10, 1832 George, his wife, Rhoda, and seven children embarked at Bristol on the ship *Emily*. Due to a cholera outbreak the ship was obliged to land at Staten Island. Once able to proceed, George and his family traveled to Cleveland, Ohio via the Hudson River and to Buffalo by way of the Erie Canal, then by lake to their final destination. Rabbitts moved to Strongsville in Cuyahoga County, where he purchased land and farmed until his death in 1849. Rhoda Nuth Rabbitts, daughter of William Nuth of Somersetshire, England, spent her later years with her daughter, Anna, in Lancaster, Ohio, where she eventually died on June 11, 1868. Charles was born to George and Rhoda Rabbitts on a farm near Herningsham, Wiltshire, England on September 2, 1820, where he remained until the age of 12, when the family came to America. At the age of 22 he moved to Newark (Licking County), Ohio, where he worked in a woolen mill and commenced learning the trades of weaving and manufacturing until 1846. He then moved to Springfield and leased a water right on Barnett's hydraulic, erecting a woolen mill on Buck Creek at Warder Street with his brother-in-law, L.H. Olds.

Importantly, the mill was the first of its kind in this part of the country. The woolen mills became quite popular and, at various times, became associated with Foos & Steele in their operations. Throughout the Midwest the mills were known for their various brands of "Rabbitts Jeans and Yarns."

On May 3, 1848 Rabbitts wed Margaret Robison, daughter of James and Margaret Wilson Robison of Wooster (Wayne County), Ohio. Six children were born (four sons and two daughters), with the eldest, Horatio, dying in infancy. Their names were James H., William S., Anna, Mary, and Charles R.

Upon his retirement from the woolen manufacturing business, Rabbitts ventured into real estate through ownership and development of city property. He was also involved with the Ferncliff Cemetery Association in selecting the site for its beautiful grounds. Rabbitts served on the school board for several terms and by mayoral appointment was a trustee of the Mitchell-Thomas Hospital. He held membership on one of Second National Bank's first boards of directors and was active in Second Presbyterian Church.

Charles Rabbitts died on December 16, 1900, with Margaret following on July 2, 1903.

James H. Rabbitts, oldest son of Charles and Margaret, was born on April 1, 1853 in Springfield, where, after several years of public schooling, he attended the preparatory department of Greenway Institute. He also worked in his father's mill, absorbing the trade of wool spinning. He entered Wooster University in Wooster, Ohio and was a member of the institution's first graduating class in 1874. He studied law in the offices of General J. Warren Keifer and Charles R. White and was admitted to the bar by the Ohio Supreme Court in December, 1876. He became a junior partner in the law firm of Keifer, White and Rabbitts until 1881, when he was chosen as the Republican candidate for Clerk of Common Pleas and re-elected in 1884 and 1887.

Rabbitts chaired the GOP's Clark County Central Committee from 1882-1890. After resigning as Clerk of Courts in 1890, Rabbitts became managing editor of the *Springfield Daily Republic-Times* for eight years. He was then appointed postmaster on May 1, 1898 by President William McKinley. He was reappointed in 1902 and again in 1906, this time by President Roosevelt. Of note, the rural delivery system was established during his incumbency.

Rabbits was elected to the Merchants' and Mechanics' Building and Loan Association's board of directors in 1898 and later served as its president. Fraternally, he was affiliated with several lodges: Anthony Lodge #455, Free and Accepted Masons, Springfield Chapter, Royal Arch Masons and Palestine Commandery, Knights Templar, Red Star Lodge, Knights of Pythias, Phi Kappa fraternity and the Literary Club of Springfield.

Rabbitts wed Cornelia Burt, daughter of the late Reverend Nathaniel C. Burt, D.D., and Rebecca A. Belden Burt, on December 7, 1882. Dr. Burt was for many years pastor of First Presbyterian Church of Springfield. He eventually traveled to Europe due to failing health and died in Rome, Italy in 1874. His remains were laid to rest there, beside those of the poet Keats. Three children were born to the Rabbitts: Margaret (who died in infancy), Burt (a young attorney of prominence who died on September 24, 1918 in his 34th year), and Frances (engaged in war work in Europe and married Captain Alden-Wilde, of the British army, in Paris).

Rabbitts died of complications from a heart attack on June 29, 1919. Cornelia lived until October 22, 1934. They rest in Ferncliff Cemetery, Section A, Lot 36 alongside other family members.

Honorable John H. Thomas
(October 4, 1826—January 23, 1901)

The Thomas family hailed from Maryland, with John Thomas born in Middleton on October 4, 1826 to Jacob and Sophia Bowlus Thomas. He was well educated in his youth, entering Marshall College in Mercersburg, Pennsylvania and graduating in 1849. He studied law with Hon. S. W. Andrews of Columbus, Ohio and completed his coursework in Springfield with Hon. William White upon his arrival in 1851.

At the end of his official term he ceased his law practice to enter manufacturing in partnership with Phineas P. Mast, founding the well-known agricultural implement company Thomas & Mast. He remained with the firm until 1872. Two years later, he resumed his manufacturing interests by organizing the Thomas Manufacturing Company with his two sons, William S. and Findley B Thomas. The company produced various agricultural implements and became one of Springfield's leading industrial businesses.

John Thomas also contributed much of his energy to city improvements, serving on city council, the Snyder Park board, and in other, similar capacities. Due to his wealth and influence he was a liberal supporter of charitable and philanthropic agencies. Perhaps his most outstanding contribution was a financial gift from John Thomas and Ross Mitchell that helped establish and maintain the Mitchell-Thomas Hospital on East Main Street. Active members of First Presbyterian Church, they supported the church in all charitable causes throughout their lives.

Thomas wed Mary Bosner, youngest child of Hon. Jacob Bosner of Chillicothe, on October 23, 1854. The couple had two sons and two daughters: William S., Findley B., Nellie (wife of Judge A.N. Summers), and Mabel (wife of L.P. Matthews).

Mr. Thomas died on January 23, 1901. The *Springfield Weekly Republic* of January 24 states, "Paralysis ends the career of Honorable J. H. Thomas Wednesday afternoon. Death came with startling suddenness ... to one of the most prominent and respected pioneer families." John Thomas holds the distinction of having the tallest monument in Ferncliff Cemetery.

Mary Thomas, in failing health after her husband's death, died in her East High Street residence on July 4, 1901. They are interred with family in Section D, lot 6 of Ferncliff Cemetery.

John Ambler Shipman
Fanny E. Grant
(August 11, 1829 - June 2, 1901)
(February 6, 1841 - September 6, 1896)

Originally from Elizabeth, New Jersey, the Shipman family moved westward and settled in Kentucky, where Clark Shipman, father of John, grew to adulthood. With an older brother living in the village of Springfield, Clark arrived in 1815 and established himself in a tailoring business known as Shipman & Company. A thoroughly enterprising young man, he reached the rank of brigadier general in the State Militia, maintaining that title until his untimely death.

On August 24, 1815 Clark Shipman married Ruth Ambler, daughter of John and Anne Brewster Hunt Ambler and born on October 30, 1799 in Bridgeton, New Jersey. Of the couple's five children, two died in infancy. After her husband's death in November of 1830, Ruth assumed the task of caring for their remaining children. She lived an exemplary life in all respects, a true pioneer from her 1808 arrival in Springfield until her death on March 29, 1883 at the age of 83. Of note, Ruth's mother, Ann Brewster, was a great granddaughter of Elder William Brewster, who traveled to America aboard the Mayflower.

Their oldest living child, Anna, died while still a young girl in 1837. Joseph Warren grew to adulthood and married Sarah Ann Hawken on July 4, 1837. His life was short-lived, however, and he died in 1841. *The Springfield Weekly Republic* of August 6, 1841 reported, "Died in this village on 18th of July of consumption, Dr. Warren Shipman, in the 25th year of his age. He left a mother, wife and infant children, and a numerous circle of friends to lament his early summons to the land of the spirit."

John Ambler Shipman was born in Springfield on August 11, 1829 and was just one year old when his father died. He received a good education given the times, participating in school until the age of 15, when he was apprenticed to learn the cabinet-making trade. He later opened a furniture store on Main Street, and due to tremendous growth, within one year built a commodious business on South Fountain Avenue between Main and High streets, where he worked for many years.

John later sold his business and accepted a position with Foos & Mulligan Furniture Company until 1877 when, upon recommendation of General J. Warren Keifer and Senator Sherman, he was appointed postmaster by President Ulysses S. Grant. He was later reappointed by President Rutherford B. Hayes and held the office until his retirement in 1885.

On December 23, 1859 John married Fanny E. Grant, daughter of early Springfield pioneers William and Nancy McCormick Grant. They were blessed with three children: Warren, Annie and Earl, all of whom received a good education and grew to maturity.

Fanny became ill and died on September 6, 1896 at the age of 54. John followed on June 2, 1901 at the age of 71. They are interred in Ferncliff Cemetery, Section I, Lot 141.

Colonel Charles Anthony
Olive E. Seitz
(September 12, 1845 - October 17, 1902)
(1850 - August 14, 1917)

Colonel Charles Anthony was one of seven children born to General Charles and Mary E. Hulsey Anthony. His early years were spent mastering an education. During the Civil War he enlisted in Company B, 86th Ohio Volunteer Infantry and was sent to Chambersburg, West Virginia for three months. Anthony re-enlisted in Company C (Montjoy's) 129th Ohio Volunteer Infantry. During the fall of 1863 as a second lieutenant, he became engaged in the Cumberland Gap and Clinch River battles.

Anthony married Olive E. Seitz, daughter of Jacob and Mary A. Stineberger, in 1866 and the couple had five children. The family settled on a farm near Urbana, and while there, served for several years, Anthony served as a member of the Champaign County Republican Committee.

Once the Champion City Guard was formed in 1873, Anthony was promoted to second lieutenant. During his term of enlistment he served as first lieutenant and captain, and later earned the rank of colonel in the Third Regiment of the Ohio National Guard. Upon outbreak of the Spanish-American War Colonel Anthony accompanied a regiment to Fernandina, Florida, where troops encamped during the battle.

Upon concluding his military service Anthony worked for Black & Anthony's dry goods store until failing health forced his retirement. His son, Theodore S. Anthony, was affiliated with the same business, while another son, Frank, labored for the Superior Drill Company. Anthony's three remaining children, Howard, Louise and Rachel, were not of age at the time of his death at age 57 on October 17, 1902. His wife out-lived him by 15 years, dying on August 14, 1917. They rest in Section H, Lot 38 of Ferncliff Cemetery.

Asa Smith Bushnell
Ellen Ludlow
(September 16, 1834 - January 15, 1904)
(August 18, 1836 - October 26, 1918)

Asa S. Bushnell was born in Rome, Oneida County, New York. The family moved to Cincinnati, Ohio and by 1851 settled in Springfield.

At the age of 17 Bushnell was clerking in a dry goods store. He later took employment as a bookkeeper and traveling salesman for Warder-Brokaw-Child, manufacturers of mowers and reapers. On September 17, 1857 Asa married Ellen Ludlow, daughter of Dr. John and Elmira Getman Ludlow, and the couple produced three children.

Along with his father-in-law, Bushnell launched a second career in the pharmaceutical business, which he maintained until the start of the Civil War. He served as Captain of Co. E. #152 in the Ohio Volunteer Infantry and also served in the Shenandoah Valley Campaign of 1864. After the war he helped organize the Springfield Gas Works, served as president of First National Bank, and became politically involved with state and local government. Bushnell was named chairman of the State Executive Committee and appointed quartermaster general for Governor J.B. Foraker. In 1895 he was nominated for Ohio's governorship and later inaugurated on January 13, 1896 as the state's 40th governor. He was re-elected in 1897 and served until 1900.

According to the Ohio Historical Society, notable achievements during Bushnell's governorship included legislation that allowed state authorities to regulate the employment of minors in factories and mines, as well as the hours they worked. Sanitary standards were issued for bakeries, and a board of medical registration and qualification was established to regulate the practice of medicine throughout Ohio.

Ellen Ludlow Bushnell was also engaged in civic and social affairs, and was of great assistance to her husband during his terms as Ohio governor. A life member of Christ Episcopal Church, she took a deep interest in its activities. She also helped organize the Lagonda Chapter of the Daughters of the American Revolution (DAR) and served as its first regent.

Bushnell died of complications of apoplexy (an early term for "stroke") three days after being stricken aboard a train en route to Springfield from Columbus, where he'd attended inauguration ceremonies for Ohio's newly elected governor, Myron Timothy Herrick. He was 69 years old.

Following her husband's death, Ellen Bushnell maintained an active life until she became ill during a trip to New York. She was later admitted to Kelly Hospital in Baltimore, Maryland, where she died on October 26, 1918.

Governor Bushnell and his family rest inside the Bushnell Mausoleum, Section P, Lot 4 of Ferncliff Cemetery.

Oliver Smith Kelly
Ruth Ann Peck
(December 23, 1824 - April 11, 1904)
(December 24, 1822 - May 9, 1901)

The grandson of native Virginian and Revolutionary War hero James Kelly, Oliver S. Kelly was born in Greene Township (Clark County), Ohio to parents John and Margaret McBeth Kelly. John served in the War of 1812 and lived as a farmer upon his return. He married Margaret (Peggy) McBeth, daughter of Alexander, in April of 1818 and continued farming until his death on September 25, 1825 at age 36. Margaret lived until March 17, 1844, reaching 49 years of age.

Oliver S. Kelly worked on the family farm during his early years, later serving a three-year carpenter's apprenticeship under Joseph McIntire. He mastered the trade around 1845 and worked one year as a journeyman.

Kelly and Ruth Ann Peck, daughter of Baker W. and Ann Peck, were married December 23, 1847.

A budding entrepreneur, Kelly later formed a business partnership with J.A. Anderson and together they launched Anderson & Kelly, considered the leading building and contractor firm of their day. Kelly later journeyed to California in 1852 to explore its vast mining regions and by 1856, returned to Ohio with a considerable fortune. He soon partnered with Whiteley and Fassler in the manufacture of reapers, purchased an old factory in 1881 and rebuilt the property as Kelly's Arcade & Hotel. A venturesome sort, Kelly purchased several other companies as well, one of which became O.S. Kelly Piano Plate Company.

Kelly was elected to city council in 1863 and served for six years. Later, he was appointed a waterworks trustee and, in 1886, while serving on the Ferncliff Cemetery board, unknowingly ensured his place in local history by constructing and fully funding a beautiful lake on the cemetery property that remains a scenic focal point to this day. Appropriately, this serene spot was christened Kelly Lake on February 23, 1887—the same year he was elected Springfield mayor.

Kelly achieved rare prominence in business, philanthropic and civic endeavors, serving in countless charitable and leadership capacities until his death on April 11, 1904 at the age of 79. Ruth Ann Kelly, 78, preceded him on May 9, 1901.

Sons Oliver Warren and Edward S. Kelly managed the family's business affairs after their father's death.

Historic O.S. Kelly Monument stands tall in Section A of Ferncliff, stately as the man for whom it's named and surrounded by members of the Kelly family.

George Gammon
Sarah 'Sallie' Bradley
(1803 - June 18, 1904)
(August 3, 1902)

The parents of George Gammon, John and Rebecca McColloch Gammon, were married on May 15, 1807 in Champaign County. Their marriage marked the first recorded governmental record of the Gammon family within the state of Ohio, as well as the first recorded among African Americans in the area soon to be known as Clark County. Ohio's 1820 Federal Census established a line of census data on the family that continued with the 1920 Federal Census Population Schedules for Champaign and Clark Counties.

Listed as charter members of the St. Paul African Methodist Episcopal (A.M.E.) Church founded in Urbana in March of 1824, John and Rebecca could not be found in any formal records associated with the Underground Railroad (UGRR). Nevertheless, it is highly likely that the Gammons were involved in UGRR activities, given that the families of Lewis Adams and Francis Reno, also charter members of St. Paul, were recorded as conductors/operators in Wilbur H. Siebert's work, *Mysteries of the Ohio Underground Railroad* and other associated documents.

David Thackery, an Urbana native and noted researcher and archivist with the Newberry Library in Chicago, Illinois, has stated that "from

1778 to 1830 the UGRR was fully operational in 14 northern states."

George Gammon (1803-1904) first appeared as a head of household with his wife, the former Sarah "Sallie" Bradley from Ross County, Ohio (listed incorrectly in census data as the "Gammrel" family). For reasons unknown, the Gammons were recorded variously as "Gamble," "Gammrel," and "Gamvelle." Cross-referencing records pertaining to the Gammon children, however, reveals that despite the surname variations, they were likely the same family.

George and Sallie's deaths were recorded in Clark County Probate Court Death Books, with Sarah passing on August 3, 1902 and George on June 18, 1904. George's involvement in UGRR activities was documented in *The History of the Underground Railroad: American Mysteries and Daughters of Jerusalem*, by Thomas Burton, M.D., published in 1925 in Springfield, Ohio. There has been historic mention of an "old colored sawyer" in D.S. Morrow's account of the local UGRR, *Yester Year in Springfield*. Listed most frequently as a laborer, George was more likely a carpenter.

A strong sense of comradeship existed among the local African-American community during the years 1850 to 1865, as evidenced by the numerous marriages that took place among families noted for Underground Railroad activity. African Americans inter-related by marriage include the Nutter, Bell, Pile, Gammon, Thomas and Durgan families, all having documented UGRR involvement.

The Underground Railroad's necessary secrecy explained the scarcity of "formal" records during that time, as the Fugitive Slave Law of 1850 made harboring "runaway slaves" a Federal crime subject to fines and imprisonment. However, recollections of the Gammon children, and in particular, Cornelia, preserved the history of the era. Their mother, "Aunt Sallie," fed and clothed runaways at the family's Piqua Place home. Meetings and "trains" operated by Abe Nutter scheduled "pickups" at the Gammon House, for transport to Urbana and Marysville. News of "slave catchers" invading the area was passed from the Robert Piles barbershop to the Hotel Porter. In this sense, UGRR activity proved a model of cooperation and trust between the community's anti-slavery segments. All races, religions and social classes worked so well together that there were no "reported captures" within the Springfield area.

The Gammon House, constructed by George Gammon, today remains one of the few existing African American "safe houses" in Ohio, if not in America, and serves as an everlasting symbol of freedom and humanity of which Springfield and Clark County can be proud.

George O.C. Frankenberg
Lucinda A. Armstrong
(November 3, 1817 - July 31, 1905)
(August 4, 1824 - January 30, 1911)

Born in Rothenburg, Germany in 1817, a teen-age George Frankenberg traveled with his family to America and settled in Columbus, Ohio, where he obtained work as a clerk. He moved to Springfield and found work clerking for John

Murdock in a dry goods store. Later identified with the grocery business, he became a member of the Frankenberg & Muzzy wholesale and retail grocery dealers. By 1856 the establishment was enlarged and renamed Frankenberg & Houghton, and was considered by many a first-class establishment.

George, whose surname has also been recorded as "Frankenburg," married Lucinda Armstrong, daughter of Oliver and Lucinda Paige Muzzy Armstrong, on September 17, 1845 and the couple had six children. They joined First Presbyterian Church before it united with Second Presbyterian Church in 1861.

Early in the Civil War when the Clark County regiment of the 44th Ohio Volunteer Infantry was organized on the Springfield fairgrounds, George Frankenberg accompanied troops to the front as its Regimental Sutler and remained to war's end.

George died following a brief illness on July 31, 1905 at the home of his son, Lucius M. Frankenberg. *The Springfield Republican Gazette* of July 31, 1905 states that he "was everybody's friend and probably never had an enemy in the world, and (he) died as he lived."

Saddened but undaunted, Lucinda continued her numerous philanthropic activities, serving as both an active missionary worker who served Springfield's poor and a teacher within the primary department of the church. She gained prominence as a temperance worker, was known as one of the original bands of "crusaders," and was active in the W.C.T.U. During the Civil War she was among the women who furnished aid to Clark County soldiers active on the front lines.

Mrs. Frankenberg was also an essayist of unique ability and wrote a number of sketches on current and local topics.

Lucinda Frankenberg died of old age and general debility on January 30, 1911, having surpassed her 86th birthday. She is buried among family in Section A, Lot 46 of Ferncliff. Of her passing, the *Springfield Daily Times* wrote on the following day, "Beloved Woman Succumbs - Noted for Kindly Acts - Woman of Rare and Lovable Personality."

John Dick
Catherine Fitzsimmons
(January 14, 1834 - November 17, 1906)
(1839 - October 17, 1879)

John Dick was born to David C. and Jessie Dick in Ayrshire, Scotland. As a young man he graduated from the Royal Botanical Gardens in Edinburgh, devoting his early years to what would become a lifelong passion of landscape gardening. Mr. Dick and his family immigrated to America in 1854, spending several initial years in New York. He later moved to Cincinnati, where he worked under the guidance of noted landscape gardener Adolph Strauch, the man responsible for America's rural cemetery movement.

Mr. Dick arrived in Springfield in the fall of 1863 and took immediate charge of Ferncliff Cemetery grounds. During his devoted and highly productive 43 years with the cemetery, Ferncliff became widely recognized as one of Ohio's most beautiful places. He married Catherine Fitzsimmons that same year and the couple had four children: Charles D, Jessie, Mary E. and

James. Sadly, the month of October proved tragic for the young family. Daughter Mary died on October 20, 1878 at the age of seven. Son Charles, age 14, soon followed, dying 10 days later on October 30, 1878. Both siblings were likely victims of diphtheria. Despondent and inconsolable, Dick's wife took her own life the following year, on October 17, 1879, as the anniversaries of her children's deaths approached.

A mere five weeks passed before Dick's father also died at the Ferncliff superintendent's residence on November 23, 1879. Likewise, his mother passed away there on July 8, 1884.

Ironically, pervasive sickness and the deaths of Dick's children had prompted Cemetery Association president John Ludlow, on August 26, 1879, to openly question the safety of the superintendent's creek-side home, concluding, "a residence on the higher part of the grounds and further away from the dampening influence of water might ... be more conducive to their health." A man of compassion and principle, Ludlow willed trustees shortly before his death to take action and arrange for proper examination of the home by qualified professionals.

Curiously, more than two decades passed before the board revisited the need for a new superintendent's dwelling during an April, 1888 meeting. The project took several more years to complete.

Following the death of his second wife, Margaret Simon, in 1903, John carried on alone, remaining in Ferncliff-property housing until his own passing on November 17, 1906 at the age of 72. The family is buried in Section B, Lots 20 and 26.

William S. Gladfelter
(1841 - 1907)

Clark County history would not be complete without mention of William Gladfelter, who arrived in Springfield in 1866 and became one of the city's leading carpenter contractors. A native of York County, Pennsylvania, William was born on a farm near Little York on March 10, 1841—one of 13 children of John C. and Louise Smith Gladfelter. John C., a well-known contractor and builder, was also a York County native and maintained a business there his entire life (the exception being a short period of time when the family moved to Guernsey County, Ohio).

William pursued his early education in local schools during the winter months and worked on the family farm in the summer. For two winters he attended the academy at York, then began helping his father as a carpenter. He traveled to Ohio in 1862 and made his first stop in Columbus, where he labored as a farm hand and shipping clerk. He arrived in Clark County in 1866, worked as a farm hand for a short time, and assisted in the construction of a park bridge in Springfield. His brother, George (and possibly John) joined him about the same time.

By 1868 Gladfelter was firmly established in the city, employed as a salesman by W.W. Diehl before embarking upon his own flour and feed store business. He also began contracting and erecting houses, building homes on property and vacant lots he'd purchased. During the next 15 years he constructed, on average, 50 houses per year, and later, as many as 40 houses annually.

Brinkman provided his own architect work, drawing all plans and specifications.

In May of 1869 William married Catherine Lankenau (1844-1908), daughter of shoemaker John D. Lankenau who'd died of cholera 20 years earlier. The couple had one son, Charles Frederick, born in July of 1872. He attended public school and graduated from Wittenberg College in 1894. His work as an auditor and bookkeeper took him to Chicago and, later, Baltimore, Maryland. He married Springfield resident Emma Town, a well-known school teacher, and the pair adopted a four-year-old son, Ryburn Barton (a nephew who became a member of the household following his mother's death). Ryburn later took over the family business for his ailing father.

Mr. and Mrs. Gladfelter were active members of the Second English Lutheran Church, having served as officers at various times. In 1892 William constructed their last residence at 618 West North Street, where they remained until their deaths. William died in February of 1907 while visiting their son in Baltimore. Catherine also died at their son's home in January of 1908. They rest in peace with other family members inside the Gladfelter Mausoleum, Section Q, Lots 6-7 in Ferncliff.

Eliza Daniels Stewart
(April 25, 1816 - August 6, 1908)

Ohio native Eliza Daniels was born in Piketon on April 25, 1816. She hailed from the Baldwin-Guthery line of ancestry, as one of her grandfathers was Colonel John Guthery of Revolutionary War fame. Tragically, both of Eliza's parents died before she reached her 12th birthday.

In 1833, while keeping house for her brother (a postmaster), Eliza was sworn in as his assistant under the administration of General Jackson, becoming, so far as is known, the first woman to ever hold such a position.

Eliza Daniels married Hiram Stewart in 1848 and gave birth to five children, but all died in infancy. Despite her overwhelming sorrow, she continued to care for her two step-sons, whom she trained and educated. In 1853 Eliza joined the Good Templers and became active in temperance work—a cause and a passion that remained strong throughout her entire life.

Stewart began her public service by participating in Soldiers' Aid Societies and the United States Sanitary Commission (later known as the Red Cross) during the American Civil War, working as a nurse and helping to organize soldiers' treatment by providing them with such necessary staples as medicine, blankets, food and supplies.

Fondly known as "Mother Stewart," Eliza was one of the original members of the Woman's Temperance Association formed in January of 1874 and later known as the Woman's Christian Temperance Union. At a time when women enjoyed no direct political power, she transformed herself into a savvy and influential crusader, drawing large crowds to her group's demonstrations by organizing parades, prayer vigils, hymn singing, petition campaigns and protest marches throughout various cities.

Eliza led the "Women's Whisky War" against saloon owners and hosted statewide temperance conventions in 1874 and 1877, and was instrumental in forming the British WCTU. "Mother Stewart" also authored at least two publications: *Memories of the Crusade* and *The Crusader in Great Britain*. Although much of her energy was devoted to the suppression of alcohol, humanitarian works sparked her interest as well. She was one of the original members of the Congress of Penitentiaries and Reformatory Discipline, organized in Cincinnati in 1870.

Roughly five years before her death, "Mother Stewart" moved to Hicksville, Ohio, where she resided with a dear friend, Mrs. Farnsworth, until her death on August 6, 1908 at the age of 92. A beautiful, gray marble monument marks the grave of "Mother Eliza D. Stewart" in Section L, Lot 570 of Ferncliff Cemetery.

William Needham Whiteley
Mary McDermott
(1834 - 1911)
(1846 - 1917)

Andrew Whiteley, father of William, was born in Harrison County, Kentucky on May 31, 1812. The family was living east of Springfield by 1814, and Andrew's early years were spent on the farm and attending school. He married Nancy C. Nelson on September 26, 1833 in Clark County, Ohio and raised a family of six children. Andrew farmed until 1852, and in the five years that followed, worked with his son to invent reaping and mowing machines and register patents. Andrew and Nancy lived a full life and are interred in Section G of Ferncliff Cemetery.

William Whiteley, son of Andrew and Nancy Nelson Whiteley, was born near Springfield in 1834. He spent his early years on the family farm, and at a young age, showed mechanical talent and an inventive mind. He apprenticed as a mechanic and became interested in harvesting machinery. From 1852 to 1855 he experimented until producing the first prototype of what would later become the Champion Combined Reaper and Mower with Sweep Rake.

According to the Smithsonian Institution Press, his "Champion Binder Company, founded in 1852, (became) one of the largest producers of farm machinery in the late 19th century."

Whiteley began to manufacture his latest inventions, joining Jerome Fassler in Whiteley & Fassler to hand-produce Champion machines. Within 12 months, O.S. Kelly, a highly skilled mechanic, joined the company under the name Whiteley, Fassler & Kelly—a business association maintained for over 25 years.

A *Champion Reaping and Mowing Machines Brochure* authored by Warder, Mitchell & Co. in 1885 and preserved in the Clark County (Ohio) Public Library claimed that the company offered "the most reliable harvester in America."

According to the library's written display information that accompanies the historic brochure, during the early 1850s "William N. Whiteley invented the first prototype of what would become the Champion Combined Reaper and Mower with Sweep Rake. This product became the foundation of an agricultural empire and Springfield became known as the 'Champion City.' By the 1870s the phenomenal success of the Champion Reaper and Mower had ushered in a golden age of manufacturing in Springfield. The demand for this product was so large that the firm Whiteley, Fassler & Kelly moved their operations to the East Street shops, one of the largest manufacturing operations in the world at that time, which covered 54 acres and employed 2,000 workers. The production of the Champion line was eventually taken over by the firm of Warder, Bushnell & Glessner, which was one of the five firms that formed the International Harvester Company in 1902."

Interestingly, around the same time that the East Street shops were formed, the Knights of Labor began organizing protests and fought for workers' rights. Additional grief befell Mr. Whiteley when he signed some ill-fated financial papers with a Cincinnati banker who later went bankrupt, causing the East Street shops to go into receivership. Sadly, they were eventually sold to pay off the subsequent debt.

For reasons still unknown, the East Street plant later burned to the ground, a devastating blow to Whiteley. He left Springfield for Muncie, Indiana, where he once again entered the manufacturing business, this time with his brother, Amos. After several years he returned to Springfield and devoted himself entirely to his inventions.

Today, Whiteley's reaper, "is one of several thousand patent models the Smithsonian museum acquired from the Patent Office in 1926," according to the Smithsonian Institution Press. Patent #16,131, entitled "Harvester Rake," was issued to Whiteley on November 25, 1856."

Ran Raider, a patent and trademark reference specialist associated with both Wright State University and Paul Lawrence Dunbar Library, complied *Dayton, Miami Valley Inventors and Inventions*. His research shows that Whiteley patented "more than 42 types of harvesters and improvements" and also held "numerous harvester patents with his father, uncle and brother, Andrew, Abner and Amos, respectively."

Raider's research also substantiates that of the Smithsonian—that Whiteley's "Champion Machine Company, founded in 1852, was one of the largest manufacturers of farm machinery in the 19th century," and that its most famous product "was the Champion reaper and mower." McCormick Company, in particular, was among the five business entities that participated in the 1902 deal and joined forces to become International Harvester.

A family man, Whiteley married Mary McDermott, the daughter of John and Catherine Setty McDermott, and the couple had two children: Helen (1872-1890) and William N. Jr. (1876-1964). Mary died on April 7, 1917 at the couple's residence, 1103 S. Fountain Avenue, at the age of 69. William Whiteley, known worldwide as "The Reaper King," fell ill in February, 1911 and his condition made daily newspaper headlines until his death on February 7, 1911. They are interred in Section G, Lot I, where a large gray marker sits atop the Whiteley family plot. Curiously, the gravestones of William N. Sr., William N. Jr. and Mary bear the name "Whitely" ... whether by error or choice remains a mystery.

Captain E. L. Buchwalter
Clementine Berry
(June 1, 1841 - October 4, 1933)
(1843 - November 16, 1912)

Edward Lyon Buchwalter was born on a Ross County, Ohio farm on June 1, 1841 to Levi and Margaret (Lyon) Buchwalter. He attended local schools and Athens-based Ohio University until the Civil War began. In August of 1862, he became a sergeant in Company A of the 114th Ohio Volunteer Infantry and served his first engagement under General Sherman in Chickasaw Bayou. He later fought under both General Sherman and General Grant, joining combat that raged in Fort Hindman, Arkansas, Alabama and Tennessee, as well as Vicksburg, Mississippi. Twice he was wounded in battle.

According to Benjamin Prince's *A Standard History of Springfield and Clark County*, Buchwalter's executive ability and soldierly qualities resulted in his commission as first lieutenant in the 53rd United States Colored Infantry. He was promoted to captain in May of 1864, but continued his service as provost marshal at Macon, Mississippi and, later, at Meridian, where he helped organize newly freed black citizens by heading the Freedman's Aid Bureau. He received an honorable discharge on March 6, 1866.

Captain Buchwalter married Clementine Berry on September 1, 1868, a woman described in *Rockels' 20th- Century History of Springfield and Clark County, Ohio* as "a lady of education and accomplishments and of much social prominence," and Delaware, Ohio seminary graduate. Some two years after her death on November 16, 1912, Buchwalter wed Marilla Andrews, a first cousin of Berry.

At war's end, Buchwalter farmed in his native county until 1873, when he and his second wife moved to Springfield. There, he became associated with the James Leffel Company, was active in the organization of Superior Drill Company in 1883, and eventually served as its president. Upon formation of the American Seeding Company in March of 1903, Buchwalter helped address an amalgamation of manufacturing concerns and was soon elected its president—a position he held until his retirement at age 70. Buchwalter also served as president of Citizens' National Bank and remained affiliated with financial institutions. At the time of his passing, Buchwalter held the title of honorary president of Lagonda Citizens' National Bank.

He'd relinquished most of his duties and responsibilities in his final decade of life due to failed hearing—a likely war-time casualty—but continued traveling and managing his 190-acre farm until his death. An October 4, 1933 obituary described Captain Buchwalter as "a man of utmost simplicity in tastes" who "could never be prevailed to wear jewelry of any kind, not even emblems of the many organizations to which he formally belonged." Funeral services in the Buchwalter home were marked by their simplicity, as was his wish. He is buried alongside Clementine and other family members in Section O, Lot 54 of Ferncliff Cemetery.

Ross Mitchell
Catherine Ann Miller
(November 14, 1824 - October 8, 1913)
(February 20, 1827 - September 12, 1878)

Sarah Keller
(September 18, 1845 - June 12, 1909)

The eldest of eight children, Ross Mitchell was born on November 14, 1824 in Landesburg (Perry County), Pennsylvania to James Blaine and Cynthia Gowdy Mitchell—the Mitchells being of Scotch-Irish descent. The family moved in 1836 to Dayton, Ohio, where Ross Mitchell's father worked as a carpenter and builder. At age 12 Ross was assisting him in various enterprises. The Mitchells rented Shartle's Mills, west of Medway in Clark County, Ohio, in 1838 and lived there for nine years before moving to Hertzler's Mill, where Mitchell's parents died.

He married Catherine Ann Miller, daughter of Donnelsville's Casper and Susan Wirtz Miller, on October 7, 1852 and continued work at the Hertzler sawmills and general store. When Hertzler sold the mill and moved to Springfield in 1853 to enter the banking business, Ross Mitchell accompanied him. After the bank was eventually dissolved he worked as a bookkeeper at Warder, Brokaw & Childs, eventually becoming a partner in the firm of Warder, Mitchell & Company. By 1881 Mitchell suffered from failing eyesight. He spent his time traveling and dabbling in real estate, investing in Ohio and Kansas farms. He also busied himself as a stockholder, officer and director of numerous companies.

Catherine Ann Mitchell died on September 12, 1878 at the age of 51. There were five children from her marriage, three of whom lived to maturity. Ross later married Sarah, daughter of Reverend Ezra and Caroline Routzahn Keller. No children were born to this union. Sarah died on June 12, 1909 at age 63.

In 1866 Mitchell was approached by John H. Thomas to join in a local, philanthropic plan to build a hospital to treat emergencies and the poor. Through this joint effort the Mitchell-Thomas Hospital was opened on December 1, 1887. Due to a lack of space and high noise levels from nearby train tracks, the facility was closed in 1904 and a new City Hospital was constructed on York and East streets. Ross Mitchell's financial contributions also aided other community developments such as Wittenberg College, the Salvation Army, Second Lutheran Church, and the Odd Fellows Home.

Described as a kind, generous and loving man, Ross Mitchell died of pneumonia on October 8, 1913 at the age of 88. He is interred with his family in Ferncliff's Section E, Lot 15.

John W. Bookwalter
Eliza Jane Leffel
(1837 - September 26, 1915)
(1842 - April 18, 1879)

John Bookwalter was born in Indiana, one of five children of David and Susan Van Gundy Bookwalter. A native of Berks County, Pennsylvania, a young David moved with his family to Ross County, Ohio, where his father, Joseph, managed a farm. Once David reached adulthood, he relocated in 1828 to Crawfordsville, Indiana to establish and home and tend to his father's previously purchased property. In 1830 he married Susan Van Gundy, of Holland ancestry, and established a farm on Little Shawnee Creek in Fountain County, Indiana. The couple raised five children.

John remained on the farm, helping with mill operations until 1865, when he came to Springfield, Ohio to conduct business and investigate the Leffel water wheel, an improved power device being manufactured at the time. He became interested in both the water wheel ... and the daughter of James Leffel. John and Eliza Jane Leffel were subsequently married on June 29, 1865 by Reverend Titus, pastor of the Evangelical Lutheran Church. That same year, he also joined his father-in-law in business. But James Leffel died in 1866, leaving John to reorganize the Water Wheel Company under the name "James Leffel & Company."

Bookwalter's brother, Francis M., came to Springfield in 1867 to assist with a mill on Buck Creek, became a Leffel Company shipping clerk and, eventually, moved to vice president and treasurer. The company thrived for 48 years under their leadership.

In 1868 the Champion Hotel Company was born and John served on its board of directors. Within a short time, donations from firms and individuals reached $100,000 and were used to build a hotel. Formally opened in September of 1869, it was named The Lagonda House. Other local buildings erected with Bookwalter's involvement included the James Leffel Company factory, the Grand Opera House, and the Bookwalter Hotel.

Eliza Jane died on April 18, 1879, and of her passing the *Springfield Republic* wrote on April 21, 1879, "deceased lady was of high culture and very refined tastes, being herself an artist of considerable skill."

By this time, John Bookwalter had become known as a distinguished traveler, accumulating 60,000 acres of land in Kansas and Nebraska. As his business ventures prospered Bookwalter became a confirmed world traveler, collecting antiquities, native art and craftsmanship from around the globe and writing of his travels in *Canyon and Crater*, *Siberia and Central Asia*, *Rural vs. Urban*, and *If Not Silver, What?* In 1893 the *Catalogue of Objects,* loaned by Mr. John W. Bookwalter to the Cincinnati Museum Association, was published.

John W. Bookwalter won the Democratic nomination for Ohio Governor in 1881 but was defeated by Charles Foster. He continued his world travels, rarely returning to Springfield. *The Weekly Republic,* dated June 2, 1892, wrote, "Mr. John W. Bookwalter, of Ohio, is in France making an exhaustive study of French village life. He is also collecting while in Europe a library for the use of his tenants at the new town of Book walter, Nebraska, which is building on the small holdings principle."

Bookwalter was philanthropic, giving to Wittenberg College and numerous charities. *The Springfield Daily Times* headlines dated December 10, 1910 noted, "John W. Bookwalter Gives Christmas Present of Seven Thousand Dollars to Poor of Springfield. The handsome gift is only exceeded in immensity by his generosity during the recent panic, when he sent from Italy his check for five thousand dollars to aid in providing food for the hundreds who were starving."

John Bookwalter became ill and died on September 26, 1915 in San Remo, Italy, a resort on the Mediterranean Sea. He was brought to the United States and is interred in Ferncliff's Bookwalter mausoleum, at rest with his wife and several members of the Leffel family.

S. Jerome Uhl
Martha A. Phillips
(1841 - April 12, 1916)
(1846 - November 3, 1930)

Mr. Uhl was born and reared in Millersburg (Holmes County), Ohio and during his early life made Springfield, Ohio his home. As the clouds of war loomed in 1861, a young Uhl enlisted for a period of three months in Company E of the 116th Ohio Volunteer Infantry, under the command of Colonel Irving. In the fall of the same year, he re-enlisted for the war's remainder under General John F. DeCourcey. He took part in many Virginia battles, mainly, Carrick's Ford, Phillippi, Cheat Mountain Gap and Cumberland Gap. He was a prisoner of war at Vicksburg, Cumberland Gap, and Jackson, Mississippi.

After the war Uhl studied art, specializing in portrait painting. He traveled to Paris to study with Augusta Emile, Carolus Duran, and Pierre de Chavannes. His works were widely acclaimed when he exhibited at Paris Salons during the 1880s. Upon his return to Springfield, Uhl opened a studio, and his public collections were viewed throughout the state. His most well-recognized local works include portraits of Phineas P. Mast, Governor Asa Bushnell, Congressman Samuel Studebaker, Benjamin H. Warder and James Alexander Hayward. During the 1890s Uhl opened a studio in Washington, D.C., where he painted many notables. A few of his portraiture and scenery paintings remain on display in both the National Art Gallery in Washington, D.C. and the Cincinnati Art Gallery.

Jerome Uhl married Martha A.M. Phillips, daughter of Jason P. Phillips of Springfield, in October of 1873. The couple's only son, Jerome Phillips Uhl, received a good education, graduating from Wittenberg in 1899. Similarly inspired, he followed his father into the arts, and in particular, portrait painting. He later moved to New York City and became a grand opera singer.

Although Jerome Uhl Sr. considered Springfield his home, he and his wife traveled extensively and lived for many years in Washington, D.C. Their last residence was Cincinnati, Ohio, where Jerome died on April 12, 1916. Martha died in Springfield, Ohio on November 3, 1930. They are interred in Section H, Lot 162 of Ferncliff Cemetery, along with many of the Phillips family.

Mr. Uhl was a member of the Society of Washington Artists and the Washington Water

Color Club. Locally, he was a member of Anthony Lodge of F. & A.M. and Palestine Commandery, #33 of Knights Templar of Springfield.

William H. Blee
Achsa Voice Blee
(1841 - March, 1917)
(1842 - August, 1920)

William H. Blee was born just east of Cleveland, Ohio in 1841. During his youth he worked for the railroad system as a passenger conductor for Cleveland-Springfield routes, which undoubtedly drew him to Clark County in June of 1869. Blee's brother, Robert, served for some time as a road superintendent, while another brother became a Cleveland minister.

Blee began his local business career with the Springfield Breweries, later serving as president of Springfield Breweries, Limited. He was a president and member of city council for a number of years, and utilized initiative to obtain the local fire department's first engine, later christened "W.H. Blee" in his honor. Mr. Blee also held the title of Springfield Savings Society vice president and was an active member of the park board.

Throughout their remaining years, he and his wife, Achsa Voice Blee, made their home at 1002 East High Street. Following a short illness, William died in March of 1917 at the age of 72. Achsa passed away three years later, in August of 1920 at the age of 78. Both are buried in the Blee Mausoleum, Section O, in Ferncliff Cemetery.

Dr. Linus Eli Russell
Alice Zischler
(June, 1848—August 2, 1917)
(November 9, 1866—January 1, 1940)

Linus Russell was born in 1848 at Burton (Geauga County), Ohio, the son of Luther and Sarah Jane Russell. His grandfather, Luther Russell, built the first log cabin in Burton Township, where Linus received his early training from neighborhood schools. At age 12 he attended Burton Academy and, later, Hiram College, where he regularly absorbed lectures given by future U. S. President James A. Garfield.

In 1871 Russell was admitted to the Eclectic Medical Institute of Cincinnati. He graduated the following year and established a medical practice in Warren. Dr. Russell also devoted some of his time to the study of law and was admitted to the bar on September 29, 1874 in Mahoning County, where he then practiced as a attorney and counselor-at-law. His professional law knowledge was frequently utilized during his medical career whenever he officiated as a medico-legal expert in important cases.

Dr. Russell served as president of the Ohio Eclectic Medical Society in 1885 and one year later was named president of Springfield's Mitchell-Thomas Hospital. He was also tapped as president of the National Eclectic Medical Association in Nashville, Tennessee due to his excellent work.

Without question, he became know as one of the most learned men of his profession. He contributed considerable literature to various publications, with many of his papers creating widespread discussion. Among them was a heavily read and circulated piece entitled, *Immediate Amputation Regardless of Shock in Railway Injuries.*

Making full use of his inventive mind, Dr. Russell solicited the assistance of Henry Voll to design and patent a self-propelled, steam-driven, horseless carriage. Its test-drive was taken on Thanksgiving Day in 1886, with rather discouraging results caused by occasional uncontrollable speed, heightened noise and the discharge of considerable ash. A second run was taken the following evening, with dire consequences. After Russell returned to his home, the carriage caught fire and its sparks eventually destroyed both the barn and the carriage. He later turned to gasoline engines and made several models with Henry Voll.

Dr. Russell was married on November 29, 1889 to Alice Zischler, daughter of George and Rachel Zischler. Their only child, Linus E. Russell, was a manufacturer and operator at Peters & Russell, Inc. He died in 1970.

The elder Russell died from apoplexy (a stroke) on August 2, 1917, and his wife on January 1, 1940. The Russells are buried in Section O, Lot 30 of Ferncliff Cemetery.

Frank McGregor   Susan Brown

David Ross McGregor   Elizabeth W. Brown

Frank McGregor
Susan Brown
(1838 - 1920)
(1843 - 1931)

David Ross McGregor
Elizabeth W. Brown
(1851 - 1928)
(1848 - 1931)

Thomas Ross McGregor
(1836 - 1936)

The McGregors settled in Clark County more than a century ago, and since that time the family has played a prominent role in Springfield's growth and development.

Frank McGregor came to America from north Scotland with his parents, Peter and Christine Ross McGregor, at the age of 13, eventually settling in Cincinnati. As war broke out among the states, Frank enlisted and served three years, aiding campaigns of the Mississippi and the battle at Vicksburg.

On June 28, 1866 he married Susan W. Brown, daughter of Ira and Olive Wilder Brown born April 7, 1843 in Lockland, Ohio. Susan became a school teacher and taught in several districts throughout Cincinnati. She also volunteered at Covenant Presbyterian and donated her time to numerous charitable causes. Of the couple's seven children, five were born in Covedale, Ohio and two in Springfield.

After the war Frank operated a Cincinnati floral business until 1875, when the family moved to Springfield and launched a nursery operation with Frank's brother, David. Frank was quite active in civic affairs and maintained membership on the Snyder Park board until late in his life. A year before his death, he relinquished many of his duties due to failing health. He died on January 11, 1920 at the age of 81. Susan's death followed on August 4, 1931, at the age of 88.

David, brother of Frank and Thomas, was born on August 7, 1851 in Cincinnati. While still a young man he traveled to Rochester, New York to spend a year studying horticulture. Upon his return to Springfield, he and Frank launched McGregor Brothers Florists in 1876—a mail-order and wholesale flower shop. The company flourished for more than 50 years, discontinued only by the deaths of the two brothers in 1932.

On December 21, 1875, David married Elizabeth W. Brown, a Lockland, Ohio native and a sister of Mrs. Frank McGregor. Like her sister, she taught in the Cincinnati school system and gave birth to three children: Roy, Hattie and David Ross.

At the age of 77, David suffered a massive heart attack and died on December 20, 1928. At the time of his death, he held positions as president of the McGregor Brothers' floral company and vice-president of the Brain-McGregor Real Estate Company. His son, Roy, was in charge of the greenhouse and David Ross was superintendent of the McSavaney Sign Company. Elizabeth died three years later, at the age of 81, on December 18, 1931.

Thomas Ross McGregor was born in Ross Shire, Scotland on May 26, 1836. He came to America with his parents in 1851 and the family settled in Cincinnati. McGregor's father, Peter, was a railroad contractor in Scotland and helped build many of Cincinnati's principal streets. In later years he moved to Springfield and lived with his son, David, until his death in 1890.

At age 20, Thomas traveled west to Oxford, Indiana, worked on a farm, then moved to Missouri to labor on a ranch. He later returned to Indiana and enlisted in the military in April of 1861, serving for three months in Company D of the 15th Indiana Infantry. Upon returning to Oxford, he helped raise Company D of the

60th Indiana Infantry, enlisted as a first sergeant at Camp Lafayette in December of 1861, and was assigned to duty with the First Brigade, 4th Division of the 13th Army Corps. His regiment saw action in Kentucky and participated in the battle of Murfreesboro, where his unit was captured and paroled. He was ordered to Cairo, Illinois and entered the campaign to open the Mississippi River. Thomas also participated in the battle and capture of Arkansas Post, the siege of Vicksburg, and the Red River campaign. He rejoined his regiment in Louisiana and did duty in Texas, Florida, Alabama and Tennessee. In March of 1865, with the war nearly over, he was ordered back to Indianapolis and promptly discharged. He eventually returned to Cincinnati and opened a fish and game store, which he sold after just one year.

On December 24, 1868 Thomas married Mercy Ann Skillman, daughter of Elijah Skillman of Hamilton County, Ohio. He operated a Cincinnati-area farm owned by his wife and, in due time, built a beautiful new home. Shortly after she died, leaving behind no children, Thomas disposed of his property and moved with his brother to Springfield. He died on February 2, 1936 in Tampa, Florida, aged 99 years and nine months.

The McGregor brothers are buried in Ferncliff Cemetery, Section Q, Lot 1.

Dr. Isaac Kay
Clara M. Decker
(1828 - 1920)
(1832 - 1912)

Charles S. Kay
(1853 - 1928)

Dr. Clarence H. Kay
(1856 - 1930)

The Kay family of early Quaker ancestry immigrated to the United States from Yorkshire, England and settled in Pennsylvania. The son of William and Susanna Unger, Isaac was born on December 8, 1828 in the Cumberland Valley near Chambersburg, where his parents resided until an 1833 move to Bedford County. Four years later the Kays traveled to West Alexandra in Preble County, Ohio, where William Unger, just 37, died soon after on November 4, 1837.

As an eighteen-year-old Isaac began studying medicine with Dr. William Gray of Lewisburg, Ohio. His three years of coursework included two lecture courses at Columbus, Ohio's Starling Medical College, where he received his degree in February of 1849. Dr. Kay practiced in Lewisburg until May, 1853, when he arrived in Springfield.

He wed Clara M. Deckert, daughter of Samuel Deckert, in nearby Miamisburg (Montgomery County) on November 3, 1852 and the couple had two children. Dr. Kay became a member of the Clark County Medical Society in May, 1854 and for 26 years authored meeting minutes and articles as its secretary. He was appointed county infirmary physician in 1858 and remained in that capacity for a number of years prior to his 1864 election as coroner. Importantly, after the Civil War depleted a newly formed Springfield YMCA of its membership and forced its temporary closure in 1865, Dr. Kay was instrumental in organizing a fund-raising effort two years later and was subsequently elected president.

Dr. Kay achieved rare distinction as a public speaker and author of both medical and lay works. A few of his writings are preserved in the *Springfield Daily News*, March 4, 1918 edition—a piece entitled, "A Story of Two Ohio Villages, Namely Springfield and West Alexandria." His work also appeared in West Alexandria's *Twin Valley Echo* on June 24, 1914 ("Early West Alexandria and New Lexington, Ohio") and in Chambersburg, Pennsylvania's *Franklin Repository* ("From Old Letter of Pennsylvanian").

Dr. Isaac Kay died on September 20, 1920 following a five-month illness. He'd practiced medicine for more than 60 of his 91 years. Clara, 80, preceded him in death on May 3, 1912. They are interred in Section H, Lot 112 of Ferncliff Cemetery.

Charles S. Kay, elder son of Dr. Isaac and Clara Kay, was born in Miamisburg, Ohio on November 4, 1853 and arrived in Springfield as an infant. As he grew older he developed a penchant for literature, contributing to Springfield and Cincinnati newspapers. He edited several publications, including *The Weekly Transcript*, a Springfield-based newspaper, and served as associate editor of *The Leffel Mechanical News*, a trade paper published by the James Leffel & Company. He also worked for two other Springfield newspapers, *The Daily Globe* and *The Sun*, where he worked as editor and associate editor, respectively. Kay's broad-based speaking abilities on a myriad of subjects drew wide acclaim and were always much in demand. Wittenberg College conferred upon him a Doctor of Literature degree in 1922.

As a businessman, he helped organize Citizens National Bank and worked as treasurer of Superior Drill Company for two decades until its merger with The American Seeding Machine Company. He then became manager and treasurer of The Springfield Light, Heat & Power Company.

Mr. Kay wed Belle C. Gunn, daughter of Captain John C. Gunn of Lexington, Kentucky, in 1893 and the couple raised four sons and two daughters. In 1916 he became affiliated with the Commercial Club of Springfield and served as president of the Lagonda Club. He was a member of Anthony Lodge, F. and A.M., Knights Templar, and the Men's Literary Club.

Following an illness of several years, Mr. Kay died on July 20, 1928. Belle followed on March 2, 1937, having led an exemplary life as an active member of the Women's Town Club, Monday Afternoon Club and Daughters of the American Revolution. The University of Kentucky graduate also participated in numerous Red Cross activities.

Dr. Clarence H. Kay was born in Springfield on October 28, 1856. As did his father, he began to study medicine at an early age, attending Columbus Medical College and, later, Miami Medical College in Cincinnati before graduating in 1882. He practiced with his father in Springfield and continued in medicine until shortly before his death. His post-graduate work specialized in electro- and thermal-therapy. Liberally curious and educated, he was also a student of psychology. At one time, Kay presided over Mitchell-Thomas Hospital. He held memberships in the Clark County Medical Society, the Ohio State Medical Society, and the Roentgen Ray Society of America, and later devoted his time as local board of education president in 1888 and 1889. During World War II he worked as a medical examiner.

Dr. Kay married Flora Wilson, a Clark County native, on September 28, 1881. They maintained active membership in Covenant Presbyterian Church, while Flora contributed her time as a member of both the Worthington Club and Daughters of the American Revolution.

Dr. Kay died on August 24, 1930, Flora soon after on April 30, 1933. The are buried in Section H, Lot 112 of Ferncliff Cemetery.

John S. Crowell
Elvira 'Ella' C. Mangold
(January 7, 1850 - August 17, 1921)
(1853 - April 20, 1937)

John S. Crowell was born in Louisville, Kentucky on January 7, 1850, the seventh child of Mr. and Mrs. Stephen B. Crowell. He attended public schools and completed an eight-year course in just six years. At the age of 11 he worked a part-time job selling newspapers while also attending school. He later worked in the *Louisville Courier-Journal* printing office and soon became its foreman. Crowell's interest in publishing flourished as he traveled among the states. He eventually met Springfielder Phineas P. Mast and entertained ideas of publishing an agricultural journal. After moving to Springfield in 1877 he established *Farm & Fireside*. Its publishing house was for years known as Mast, Crowell & Kirkpatrick, but later became the Crowell-Collier Publishing Company. The firm purchased *The Home Companion* and changed its name to The Ladies' Home Companion. Crowell sold the company in 1966 and its offices moved to New York, but the plant remained in Springfield and continued to thrive.

At the age of 27, Crowell wed Ella C. Mangold, a product of a prominent Louisville, Kentucky family, on November 20, 1877. The couple had several children, with two daughters reaching adulthood.

Upon arriving in Springfield on August 17, 1877, Crowell became active in numerous religious, educational, charitable and business organizations. He held many positions of trust and honor: director of First National Bank of Springfield, Elwood Myers Company president, director of Columbia Life Insurance Company of Cincinnati, board of trustees president for the Western College for Women in Oxford, Ohio, and First Presbyterian Church and Covenant Presbyterian Church elder. He was also director and treasurer of the YMCA, as well as president of its board of trustees.

Crowell died at Christ Hospital in Cincinnati following a two-year battle with cancer on August 17, 1921. A few years after his death, Ella left Springfield to live with a daughter in Newtonville, Massachusetts. During her time in Springfield, Mrs. Crowell was a member of First Presbyterian Church and the Monday Afternoon Club. She was also affiliated with the Springfield Country Club and the Lagonda Club.

She died on April 20, 1937 in Newtonville and her body was returned to Springfield for burial beside her husband in Section I, Lot 55 of Ferncliff Cemetery.

Scipio Eugene Baker
Jessie Foos Baker
(1860 - 1921)
(1868 - 1949)

Margaret Evelyn Baker
(May 29, 1896 - January 10, 1986)

Ezra D. Baker, grandfather of Scipio, hailed from New Jersey and moved to Clark County in 1805. He married Anna Morgan and together they reared four children. Baker was occupied with farming and enjoyed public service. He served four terms as a county commissioner.

Alonzo Addison Baker, son of Ezra and Anna Baker and father of Scipio Baker, was born on the Baker farm in Enon, Ohio in 1831. He received an excellent education. In 1845 Baker began the study of medicine and in 1870 graduated from Ohio Medical College. He married Margaret Miller in October of 1856 and the couple raised four children. Alonzo continued his medical practice until 1878, when he redirected his energies into manufacturing. An eventual president of the

Champion Chemical Company, Baker also helped organize the Springfield Metallic Casket Company.

Scipio Eugene Baker, son of Dr. Alonzo A. and Margaret Miller Baker, was born in Donnelsville, near Enon, on June 12, 1860. As a young man, Scipio located to Springfield and graduated from Wittenberg College in 1878. He studied law for a time but gradually surrendered his aspirations and, instead, entered the business world with his father, helping found the Springfield Metallic Casket Company and eventually becoming manager.

In 1889 he disposed of some of his holdings, became interested in the salt-producing industry, reorganized, and became president of the Royal Salt Company of Kanapolis, Kansas. He also served as director of the Western Salt Company of St. Louis, Missouri.

Scipio married Jessie Foos, daughter of John and Samantha Mark Foos, on June 25, 1895. The couple had one child, Margaret.

Scipio became president and treasurer of the Foos Gas Engine Company in 1897. He was also named president and director of Springfield National Bank, as well as treasurer of Champion Chemical Company (upon his father's death). He continued his business enterprises and travels until becoming ill in Colorado Springs, Colorado, where he died at Bethel Hospital in September of 1921. Jessie died in June of 1949 at 80 years of age. They are buried together in Section P, Lot 19 of Ferncliff Cemetery.

Charles Frank McGilvray
Addie Francella Gray
(January 22, 1849 - June 26, 1922)
(1856 - December 3, 1935)

The McGilvray family came to America from the rolling hills of Scotland in 1782 and settled on a New England farm. Charles, the fourth generation in America and one of three children of Thurston and Mary Ballard McGilvray, was born on January 22, 1849 in Peterboro, New Hampshire. Two years later, his father traveled to California in search of gold and died there at the age of 39. With assistance from family members, Charles and his siblings were educated in the Peterboro public schools and worked on the McGilvray farm. Once Charles turned 18 he entered the foundry, learned the trade, and began work as a journeyman.

He was married in 1873 to Addie F. Gray, the daughter of David Gray, a Peterboro builder and contractor. The couple had one child who died at the age of two. In 1876 they moved to Cleveland, Ohio and, later, to Elmira, New York, where

Charles managed the foundry at New York State Reformatory.

The couple arrived in Springfield in 1884 and Charles was placed in charge of the Robbins & Meyers Company Foundry. By 1888 the business was incorporated, and within four years, J.A. Meyers teamed with McGilvray to buy the company outright. Mr. H.E. Meyers became a member of the company and McGilvray established himself as president. His interest in both financial matters and civic affairs made Charles' advice invaluable, and he was soon elected city commissioner and mayor.

Although McGilvray suffered with diabetes for a number of years, the disease didn't deter him from giving aid to those in need. The family was widely known for philanthropic activities and contributed greatly for the benefit of all. Charles McGilvray died on June 26, 1922 at the family home on 123 North Limestone Street. Mrs. McGilvray later moved into a suite at the Shawnee Hotel, where she remained until her death on December 3, 1935.

Prior to Charles McGilvray's death, plans had been made to dedicate the family home to the YMCA for a new building site. In 1925 a portion of the grounds that occupied the McGilvray Annex of the "Y" was donated to the organization, and by 1929, the entire home. Among Charles' hobbies was sponsoring McGilvray's Boys Club, composed largely of underprivileged boys whose YMCA memberships were paid by Mrs. McGilvray. Clark Memorial Home and the old Tuberculosis Camp were also blessed by her generosity. Wittenberg College received annuity gifts to erect the McGilvray Natatorium, which housed the swimming pool in the campus' Health and Physical Education Building. Other known contributions were to Christ Episcopal Church, where the McGilvrays were members, as well as Oakland Presbyterian, Central M.E. Church and numerous other institutions, families and individuals.

Having devoted their lives to philanthropy, Charles and Addie rest in peace inside the McGilvray Mausoleum, Section R, Lot 339 in Ferncliff.

Sergeant James C. Walker
(November 30, 1843 - April 8, 1923)

James C. Walker was born on November 30, 1843 in Harmony (Clark County), Ohio, about six miles east of Springfield. He moved with his parents to Springfield in 1857 and worked as a carpenter shop apprentice. As the Civil War broke out, Walker enlisted in the first company, afterward known as Company K, 31st O.V.I. At the end of his three-year service he re-enlisted at

Chattanooga, serving in every battle the Army of Cumberland fought, roughly 81 engagements. He received several commendations for his bravery, the highest being the Congressional Medal of Honor.

Walker mustered out on July 20, 1865 and returned to Springfield. He then spent about eight years participating in the construction of the Vandalia railroad in Effingham, Illinois. Once again returning to Springfield, Walker married Susie P. Llewellyn and, over the years, the two were blessed with several children.

Walker labored in factories until his appointment to the police force in 1883. He was placed in the station house and drove a wagon until being named chief on November 25, 1885 at the age of 41. Walker received a salary of $1,000 and from that paid all incidental expenses. As chief of police, he was in charge of 18 patrolmen. Four worked a day shift, four staffed the patrol house, and 10 worked at night. Notably, Walker started the first rogue's gallery at his own expense. Commenting to press of his day on his career as chief, Walker vowed to "get rid of all the bad characters by scaring them to death" and to deal swiftly with "bums on the trains." He quickly developed a reputation for maintaining efficiency within his small department.

James Walker died in Springfield on April 8, 1923 at age 79. At the time of his death, he was survived by a wife, two daughters and three sons. He is buried in Ferncliff Cemetery, Section I, Lot 117.

Henry L. Schaefer
Bertha Orthman
(July 31, 1850 - January 7, 1924)
(March 21, 1851 - February 19, 1930)

Leonard Schaefer, a young man of 26 years, immigrated to the United States from Reichenbach, Wuerttenberg, Germany, arriving in Springfield in June of 1849. His original intention was to settle in Cincinnati, where there was a large German population, but a cholera epidemic made travel unwise. He instead took rooms at the American House. Soon after, the dreaded scourge invaded the city and many citizens fell victim. Schaefer quickly found work as a locksmith, a trade he'd learned in Stuttgart before coming to America.

On August 12, 1849, Reverend Chandler Robbins married Leonard and Rosina D. Esslinger, to whom he had been engaged in Germany. The couple had four boys and two girls. After nearly 20 years of marriage, Rosina died on June 11, 1869. Leonard later met and married Bertha Kurz, a relative of his late first wife, during a trip to Germany. Four sons were born to the pair.

At the time of Leonard Schaefer's death on May 5, 1895 at the age of 72, the local newspaper described him as "one of Springfield's most highly respected citizens and one of the oldest German residents of this city." Henry L. Schaefer was the first-born of Leonard and Bertha Schaefer's children. He arrived on July 31, 1850 in Springfield and, as a young man, attended local public schools. He traveled with his father to Germany in 1869 and attended two terms in the Government Technical College at Stuttgart, taking a course in mechanical drawing. Henry worked in his father's machine shop after returning to Springfield, and later, as a foreman in the Champion Bar & Knife Company tool department. He traveled to Chicago in 1895 and took courses in embalming. He established a mortuary business at 226 W. Main, later moving Henry L. Schaefer & Son to 736 S. Limestone Street.

Schaefer was married to Bertha Orthmann, a Cincinnati native born on March 21, 1851 and the daughter of Dr. and Mrs. Fred Orthmann of Hillsboro, Ohio. The two were wed on July 30, 1872 and produced three children: Katherine, Bertha, and Theodore.

Henry Schaefer was quite prominent in city business affairs, serving two terms as coroner. His first commission bore an historic signature, that of William McKinley, Ohio's governor at the time and, later, America's 25th president. His second term was commissioned by Governor Asa S. Bushnell. He later served on the Springfield Board of Education, representing the 6th Ward during the 1880s. He was again elected in 1904 during the board's first at-large election—a position he held until 1920. His years of dedicated school board service were publicly recognized with the naming of a junior high school in his honor: Henry L. Schaefer School, today known as Schaefer Middle School.

Schaefer for many years sat on the board of directors for St. John Evangelical Church, and also took part in numerous fraternal organizations. He served as secretary and treasurer of the local committee that established the Ohio Pythian Children's Home, and was a member of a committee that helped deliver the Independent Order of Odd Fellows Home to Springfield.

He belonged to Clark Lodge #101, Free and Accepted Masons; Springfield Chapter #48, Royal Arch Masons; Springfield Council #17, R. and S.M.; Palestine Commandery, #33, Knights Templar; Antioch Temple, Shrine of Dayton; the Dayton Consistory and the 33rd degree.

Schaefer founded the Order of the Eastern Star chapter in Springfield and was its first Worthy Patron. He was also a member of the Moncrieffe Lodge #33, Knights of Pythias, and held the rank of Major in the uniform rank of that order. He belonged to Goethe Lodge #334, Independent of Odd Fellows; Mad River Encampment #16 (in which he occupied all the chairs); Canton Occidental, Patriarchs Militant, in which he was a past senior captain; Champion Council #2 Juror Order United American Mechanics; Violet Council #29, Daughters of America; and the Clark County Historical Society. Schaefer was also at one time a director of both the Clark County Buildings & Savings Company and Springfield Light, Heat & Power Company.

Henry Schaefer died of liver cancer on January 7, 1924. Bertha C. Schaefer passed on February 19, 1930. Mrs. Schaefer was one of three charter members of the Ladies Aid Society of St. John's Evangelical Church and took part in the society from its founding. Likewise, she was a charter member of the Home City Chapter of the Eastern Star and belonged to the Fidelia Rebekah Lodge and Pythian Sisters.

The Schaefers are buried in Ferncliff Cemetery, Section P, Lot 18.

Florence E. Kinney

Marietta E. Kinney
Florence E. Kinney

(April 11, 1841 - January 15, 1927)
(October 11, 1869 - January 23, 1935)

The daughter of John and Elizabeth Enoch, Marietta was born on a farm in West Liberty (Logan County), Ohio. She spent her early years completing both family chores and a liberal education. After marrying James Kinney in 1862 the couple moved to Illinois for several years before eventually settling in Springfield, Ohio. The only known child of James and Marietta was Florence, born in 1869.

Mrs. Kinney was at the forefront of many Springfield-based, philanthropic and Christian causes, being among those who organized St. Paul Methodist Episcopal Church, where she taught Sunday School for 20 years. She was also named life secretary for the church's Ladies' Aid, a position she held for 33 years. Active in the formation of the Woman Crusaders, which later developed into the worldwide Woman's Christian Temperance Union, Kinney served as Springfield branch president for six years.

Through her work in the Temperance Union she met Reverend "Billy" Sunday, a famed evangelist born in Ames, Iowa on November 19, 1862. Sunday's father died of pneumonia inside a Missouri military camp, after which he and his brother, George, were raised in a Soldier's Orphanage in Glenwood, Iowa. He began a baseball career with the Chicago White Stockings in 1883 and, during an eight-year period, played for both the Pittsburgh Pirates and Philadelphia Phillies. He later quit baseball to become part of the Young Men's Christian Association, assisting another evangelist with revival meetings throughout the Midwest. In 1903 he was ordained a preacher by the Presbyterian Church. Utilizing the newly discovered wonders of radio, Sunday reached listeners throughout America, converting them with fiery sermons on the evils of alcohol and in support of Prohibition.

During Sunday's campaign in Springfield, Florence Kinney became interested in his work and joined the organization. She'd held a deep interest in the church since childhood, being a student of the Bible and a devoted member of St. Paul's Methodist Episcopal Church. She later graduated from Springfield High School and the Young Ladies' Seminary, where she eventually taught.

Florence Kinney spent the next 18 years working tirelessly for the campaign, and for many years, was the only woman member of the organization. She rose to dean of Iowa Biblical College in Waterloo, Iowa before an accidental fall atop ice took her life. She is interred alongside her mother in Ferncliff Cemetery's Section O. Many of

"Billy" Sunday's devotees attended their funerals, with Reverend Sunday speaking at Florence's service.

The skilled evangelist died of a heart attack on November 6, 1935. A memorial service was held at Moody Memorial Church in Chicago and was attended by thousands of mourners.

William R. Burnett
Mary C. Monahan
(August 17, 1846 - November 1, 1928)
(1846 - January 22, 1923)

Burnett was born to early settlers John and Mary Jones Burnett on August 17, 1846 in Clark County, Ohio. He attended Western School until the age of 14, when he entered the shops of Whitely, Fassler and Kelly and remained for three years, becoming a skilled machinist. During the Civil War he enlisted in Company A, 4th Battalion, Ohio Independent Cavalry and remained until his discharge in 1865.

Burnett married Mary Catherine Monahan, daughter of John and Eliza Monahan, in October of 1865 and raised two known sons, Theodore A. and Levi Herr. Theodore attended college in New York and entered veterinary medicine, while Levi became a lawyer.

Once his military service ended, Burnett returned to his former job in the machine shops and continued working there for 23 years. After a year of employment with Union Central Life Insurance Company of Cincinnati, he later embarked on the grocery business for a 10-year period.

Always interested in city and county politics, Burnett was thrice elected Springfield mayor. He eventually retired from government and traveled to Sandusky, Ohio, where for 12 years he served as Commandant of the Soldiers Home. More than 900 Civil War veterans resided there at the time.

Burnett held many other public offices, including six years on the school board, four years representing the first ward on the city council, two years on the board of public service, four years on the police and fire board, and four years as a government gauger at William Burns Distillery. He was also affiliated with numerous lodges and a member of the Lutheran church.

Following several months of ill health, William, fondly known as "Uncle Billy," died on November 1, 1928. He was preceded in death by his wife, who passed away in January of 1923. They are buried side by side in Ferncliff's Section P, Lot 15.

John Crabill
Barbara Ellen Ober Zimmerman
(July 5, 1847 - March 23, 1930)
(1852 - April 29, 1940)

Among Clark County's earliest pioneer families is the Crabill family, which migrated from Louden County, Virginia around 1808. At the time, Clark County was part of nearby Champaign County. With their two children and other siblings, David and his wife, Barbara Baer Crabill, settled on a farm with Crabill's two brothers, Thomas Voss and Solomon, working the soil until David was able to purchase land for himself in 1813.

During the War of 1812 David served as First Sergeant in Captain Lingle's Company, formed in Champaign County and under the command of General Wayne. After service, Crabill returned to farming his land and constructing a home for his family. Although that residence remains initially listed in 1826 county tax records, it may have been constructed much earlier. The resultant beautiful brick dwelling of Federal design was home to David, wife Barbara, the couple's 12 children and a few additional members of the next generation.

Thomas Voss Crabill, second son of David and Barbara, was born on the farm in Moorefield Township, Clark County, Ohio, on November 2, 1810. He assisted his father until his January 31, 1833 marriage to Sydney Yeazell, daughter of Abraham and Mary Curl Yeazell. He then rented one of his father's farms and later purchased the property, living there with his wife. Thomas died on September 5, 1884. Sydney's death followed on December 8, 1907. Their union produced 13 children—six boys and seven girls.

Born July 5, 1847 and the ninth child of Thomas and Sydney, John Crabill grew up on the family farm and soon specialized in stock raising. He married Barbara E. Zimmerman, daughter of Isaac and Anna Zimmerman, on December 19, 1872. The marriage produced three children: Ada C., who wed William Mahar, a local attorney, Clark R. and Pearl Preston. The family later moved to a farm at the intersection of East National and Bird roads, where John lived until his death in the spring of 1930. Barbara died 10 years later in 1940. Both are buried in Ferncliff Cemetery, Section E, Lot 104. A few of their siblings rest in Section H.

When the U.S. Army Corps of Engineers began construction of Clarence J. Brown Dam & Reservoir in 1966, the Crabill House was vacated. The Clark County Historical Society acquired the property in 1973 and began extensive renovation that lasted through 1976. The group again rehabilitated the home in 1997 and 1998. Today, the Crabill homestead is listed on the National Register of Historic Homes.

Interestingly, the Clark County Historical museum houses the Conestoga wagon used by the Crabills during their westward migration from Virginia to Ohio, as well as David Crabill's sword from the War of 1812.

General J. Warren Keifer
(January 30, 1836 - April 22, 1932)

Born on a Bethel Township (Clark County, Ohio) farm on January 30, 1836 to Joseph and Mary Smith Keifer, J. Warren spent his formative years completing farm chores and attending school. He taught at the Black Horse Tavern school on Donnelsville Creek in 1852, and the following year began the study of law, entering Antioch College in Yellow Springs as a member of the school's inaugural class. Four years later, he was a student of law in the offices of Anthony and Goode until admitted to the practice on January 12, 1858.

J. Warren married Eliza J. Stout, daughter of Springfielders Charles and Margaret Stout, on March 22, 1860 and the two eventually parented three sons and one daughter: Joseph Warren, William W., Horace C. and Margaret.

Well-known to Civil War historians throughout the United States, J. Warren Keifer was among the first group of soldiers to enlist as privates on April 19, 1861 in defense of the Union. Within two days, he was commissioned major of the Third Ohio Volunteer Infantry for a three-month period, and prior to its expiration date was commissioned for an additional three years. He fought in the battle of Rich Mountain on July 11, 1861 (and a total of 29 battles to end the war). Keifer was promoted to lieutenant colonel on February 21, 1862 and later commissioned colonel of the 110th Ohio Volunteer Infantry on September 30, 1862.

At the battle of the Wilderness on May 5, 1864, J. Warren was seriously wounded but resumed command in August despite his disability. He was appointed brigadier general by brevet for valiant services on November 30, 1864, and President Abraham Lincoln rewarded his bravery and courage by brevetting him brigadier general one month later, on December 29, 1864. During the final two years of the war Keifer commanded the brigade under General Sherman before mustering out of service on June 27, 1865. By July of that year he'd resumed the practice of law in Springfield.

In 1868 Keifer was chosen as an Antioch College trustee and continued to hold numerous positions of trust and honor at the college well into advancing age. He served as Lagonda National Bank president upon its organization in 1873 and remained in that capacity until a merger with Citizens National Bank selected him honorary president of Lagonda-Citizens National Bank.

Named commander of the Ohio Department of the Grand Army of the Republic in 1868, Keifer organized a board of control to establish a soldiers' and sailors' orphans home in Xenia and later served among its trustees.

Widely known as a tremendous orator, the politically motivated former general served in the Ohio Senate from 1867 to 1870, and in 1876 was an at-large delegate to the Republican National Convention in Cincinnati. That same year Keifer was also elected to congress (1877-1885), setting the stage for his rise to Speaker of the House on December 5, 1881—a position of power he held until March of 1883.

When war with Spain was declared in 1898, Keifer was appointed major general by President William McKinley and assumed command of the 1st Division of the 7th Army Corps in Florida, moving into Cuba with 16,000 men and establishing headquarters near Havana. Troops took control of city on January 1, 1899, but on March 12 of that year, Keifer was called home from the war by the death of his beloved wife, Eliza (1835-1899). He formally mustered out of service on May 11, 1899 and resumed his Springfield law practice.

By 1905 J. Warren was a sitting member of the 59th U.S. Congress and served until 1911. He represented congress during the International Peace Conference at Brussels, Belgium in 1910, and in 1912 was named an honorary life member of the International Peace Union in Rome, Italy,

The revered war hero and eloquent statesman, at times seemingly larger than life, died on April 22, 1932 at the grand age of 96. J. Warren Keifer is interred with family members in Section H, having rebuffed burial offers in Arlington National Cemetery for the gently sloped beauty of Ferncliff in the city he called home.

Dr. Benjamin F. Prince
Ella Sanderson
(December 12, 1840 - September 11, 1933)
(1847 - February 17, 1911)

Benjamin Franklin Prince, son of William and Sarah Nauman Prince, was born in Westville (Champaign County), Ohio on December 12, 1840. He was raised on the family farm and educated in country schools. He entered the preparatory department of Wittenberg College in 1860 and graduated in 1865. He received his M.A. in 1868, and in 1891, Wittenberg awarded him its Doctor of Philosophy degree. While Dr. Prince was a theology student for a time, he was appointed an instructor in 1866 and devoted his life's remainder to Wittenberg, serving in a variety of offices that ranged from professor to vice president.

Dr. Prince edited the book *Centennial of Springfield* in 1901 and contributed numerous articles on historical topics. He authored *The Rescue Case of 1857*, *The Influence of the Church in the Organization of Modern Europe*, *Beginnings of Lutheranism in Ohio*, and *Theological Education in Wittenberg College*.

Dr. Prince married Ella Sanderson on August 3, 1869 and they were blessed with four children. He remained active in public affairs, having served many years on the city council and as president of the Clark County Historical Society. He also held many positions at the state and national level.

Mrs. Prince was a prominent volunteer for the First Lutheran Church and served as secretary of the executive committee of the Woman's Home and Foreign Missionary Board of the Lutheran Church, which was headquartered in Philadelphia. She died on February 17, 1911. Dr. Prince passed away in 1933, having lived to age 92. They are buried in Ferncliff Cemetery, Section L.

In 1977 the Benjamin Prince Society was founded to recognize individuals willing to invest in Wittenberg University and carry on the work to which Dr. Prince dedicated his life.

William Bayley
Mary E. Discus
(July 28, 1845 - February 4, 1934)
(1850 - December 24, 1933)

WIlliam Bayley was born July 28, 1845 in Baltimore, Maryland, the son of William and Mary Ann Bayley. Mr. Bayley was reared in his native city of Baltimore and was educated by its city schools. Before the age of 20, he became a teacher at Baltimore's Maryland Institute.

In 1871 Bayley organized the Bayley-Remington Company of Wilmington, Delaware; now known as the Remington Machine Company. He married Mary E. Discus, a Baltimore native, on February 21, 1871 and the couple moved to Springfield four years later. Bayley found employment in the design field, joining Whiteley, Fassler and Kelly Manufacturers until 1889, after which he secured work with the Rogers Fence Company. Around 1904 the company reorganized under the name of the William Bayley Company, and Bayley served as president until his death. His sons, meanwhile, had taken over active management some time prior.

Bayley was a member of the first city park commissioners board and did much to improve Springfield's parks. He also designed and erected numerous bridges throughout the city and Clark County.

Bayley died on February 4, 1934 at the age of 88 following a bout with pneumonia. His wife, Mary Ellen, died on December 24, 1933 of congestive heart failure at the age of 83. They are buried in Ferncliff's Section C, Lot 31.

Newton Hamilton Fairbanks
Lucy C. Cruikshank
(December 10, 1859 - March 22, 1937)
(September 2, 1864 - July 11, 1944)

Newton Fairbanks came from New England stock. His father, Loristan Monroe Fairbanks, was born in May of 1824 in Barnard, Windsor, Vermont. His mother, Mary Adelaide Smith, was a native of Spencertown, New York. The family moved west and settled on a farm in Unionville Center (Union County), Ohio, where Loristan pursued farming and wagon-making.

Newton Fairbanks, one of 10 children, was born on the Union County farm on December 10, 1859. He attended Delaware High School and, later, Ohio Wesleyan University, graduating in 1884. An 1886 University of Cincinnati Law School graduate, Fairbanks began his career in Kansas City, Missouri. After one year, he relocated to Los Angels to become general manager of Pacific Land Improvement Company, a Santa Fe Railroad subsidiary.

He moved to Chicago in 1890, where he practiced law for 10 years before arriving in Springfield to serve as vice president and treasurer of the Fairbanks Company, a piano plates manufacturer. He also presided as president of Indianapolis Switch & Frog and Mutual Fire Insurance companies. After organizing American Trust & Savings Bank in 1906, he served as its president until 1916. He also teamed with his brothers, Luther M. and Charles W., to construct the Fairbanks Building.

Politically active, Fairbanks served as a Seventh Ohio District delegate to the 1899 Republican National Convention that nominated Theodore Roosevelt and Fairbanks' brother, Charles, for president and vice president of the United States—a winning ticket. Newton later spent two years chairing the Clark County Republican Central and Executive committees and the State Republican Committee.

He served as an at-large Ohio delegate to

the 1920 GOP National Convention in Chicago, helping nominate Warren G. Harding for the presidency. He was elected Clark County Representative to the State Legislature in 1934 and re-elected in 1936—a position he held until his death from heart disease on March 22, 1937.

Fairbanks was a member of Covenant Presbyterian Church, Clark Lodge #101 F.& A.M., and Sons of the American Revolution. He also served as president of both the George Rogers Clark Chapter of Springfield and the Ohio Society (1935).

Mr. Fairbanks wed Lucy Cruikshank, daughter of Delaware, Ohio natives George H. and Augusta Smith Cruikshank, on November 17, 1887. She was born on September 2, 1864 in Delaware County, later graduating from Ohio Wesleyan University in 1883. The couple produced five children.

Mrs. Fairbanks devoted her time to Covenant Presbyterian Church activities, and served as regent for the Lagonda Chapter of the Daughters of the American Revolution from 1916 to 1918. For many years, she was also an active Y.W.C. board member. She died suddenly on July 11, 1944 from a massive cerebral hemorrhage and is buried beside her husband in Ferncliff Cemetery. More than 17 family members are interred in Section Q, Lot 2.

A *Press Republic* article dated December 25, 1902 honored the Fairbanks' stately memorial, noting, "A handsome monument has been erected on the Fairbanks lot in Ferncliff Cemetery. It is situated just across the driveway, west of the Bushnell Mausoleum. The dimensions are unusual. It is 11 feet, 4 inches by 7 feet at the base, and 8 feet, four inches high, and is of Quincy granite with (a) hammered finish. On the face of the stone, in raised letters, is the name 'Fairbanks.'

"On this lot are buried the father, two brothers, and a sister of Senator Charles W. Fairbanks, N.H. Fairbanks and Mrs. M.L. Milligan. This work of placing the monument was completed last Saturday. ... May they all rest in peace."

Ralph W. Hollenbeck
Ellen B. McGrew
(1880 - 1938)
(1882 - 1964)

The Hollenbecks date to the early 1800's in Great Barrington, Massachusetts, where Dwight Wheeler, father of Ralph, was born on April 12, 1844. Dwight received a liberal education until age 18, when he entered the wholesale flour and feed business in Great Barrington owned by his father, John Van Dusen Hollenbeck. After a few years he decided a move west might produce better business opportunities. Upon his arrival in Circleville, Dwight established a wholesale and retail clothing business with his future father-in-law, George Melvin. He married Ada A. Melvin on January 1, 1867 and their union produced seven children. Dwight remained in Circleville for the next 20 years, maintaining a successful business until moving to Springfield to pursue new opportunities. He entered the insurance field, representing Union Central Insurance Company of Cincinnati and, later, John Hancock Insurance Company. While there he became general manager for the district, a position he held until his death at age 57 on November 7, 1896.

Ralph W., son of Dwight and Ada Hollenbeck, was born in Circleville on September 19, 1880 and during his early boyhood moved to Springfield. Upon graduation from Wittenberg College (Class of 1901), he began a career with International Harvester Company, remaining there until his retirement as the local plant's executive manager. Ralph then pursued a second career with Morris Plan Bank, being named an eventual vice president and director. Teaming with Frank J. Braun and Allan McGregor, he later established Credit Life Insurance Company and served as its president until his death. He also served during that time as director of the First National Bank and Trust Company.

On June 15, 1905, Ralph Hollenbeck married Ellen B. McGrew, daughter of Mr. and Mrs. J. Frank McGrew and granddaughter of Asa S. Bushnell, Ohio's governor from 1898 - 1902. Seven children were born.

As his father before him, Ralph showed great interest in the city's welfare and pursued an active role in the YMCA and Boy Scout movements, serving as president of the Tecumseh Council Boy Scouts and as a member of the Tecumseh Area Council. He was also a member of St. Andrew's Lodge #619 and of the Ancient Accepted Scottish Rite, as well as a member of the Springfield Rotary Club and the Springfield Chamber of Commerce. Ellen, meanwhile, was equally active in civic and social circles, including the Clark Memorial Home, the Young Women's Mission, and the Springfield Country Club. The Hollenbecks were members of Christ Episcopal Church.

Following six weeks of illness, Ralph died of heart disease on July 26, 1938 at age 58. Ellen died on June 17, 1964, having lived to age 81. Both are interred in Section P, Lot 10 of Ferncliff Cemetery.

Dwight Wheeler, elder son of Ralph and Ellen Hollenbeck, was born November 21, 1909 in Springfield. A graduate of Phillips Exeter Academy, he earned a Bachelor of Arts degree at Princeton University (Class of 1932) before serving in the Navy during World War II as Quartermaster of the USS Regulus in the Pacific Theater.

Following his military service Dwight Hollenbeck entered the banking business and served as the inaugural vice president of First Morris Plan Industrial Bank. He was elected to its board of directors in 1939. First Morris eventually became Security National Bank, where

Dwight remained on the board for 57 years. He then succeeded his father and brother, John, as president of the Credit Life Insurance Company, and later became chairman of the board. He also served on the board of directors for Credit Life Companies, the holding company of Credit Life Insurance, Central Penn Insurance, and Sterling Life Insurance.

Dwight Hollenbeck married Jane Pratt Bayley, daughter of Lee and Beebe Bayley, in 1939 and the couple raised two children.

In 1987, a majority share of Credit Life was sold to Crabtree Capital Corporation. Aon Corporation later bought the business from Crabtree and relocated away from the city.

Dwight and Jane Hollenbeck were benefactors of many charities and institutions in Springfield. They gave generously to Wittenberg University, the Clark County Historical Society and Springfield Art Museum, all while making numerous other anonymous donations. Both were members of Christ Episcopal Church.

Dwight became ill in early 1996 and died on July 7 at the age of 86. A brother, John, and sister, Martha Harrison, preceded him in death. He is buried alongside family members in Section P, Lot 10 in Ferncliff Cemetery.

Max L. Kleeman
(July 5, 1867—February 3, 1940)

Max was a native of Columbus, Ohio, having been born on July 5, l867 to Louis and Clara Weixelbaum Kleeman . His parents, originally from Germany operated a jewelry business in the Columbus area. When Max was eleven years of age, the family moved to Cincinnati where his father engaged in the sale of China and dishes and where Max completed his education. He then started his business career as a buyer for a Cincinnati grain firm, traveling over the United States purchasing grain for the company. After several years with the grain company he went in partnership with W.J. Strauss and founded The Western Furniture Company.

On August 12, 1902, Mr. Kleeman married in Cincinnati to Miss Bess Willoughby of Mason, Ohio. From the time Max arrived in Springfield one year later, he became active with the civic and business circles of the city and became one of its leaders.

A prominent figure in local business and financial circles he entered the business of The People's Outfitting Company, a furniture store located at 21 S. Fountain Avenue, which had been founded by Mr. Kleeman's older brothers, Jacob and W.L. Kleeman. He was also an organizer and executive in partnership with two other brothers, Oscar and Edward and a nephew, Walter B. Kleeman, of The Associated Furniture Company, a firm controlling furniture stores in Ohio, Pennsylvania and New York. His other business affiliations included serving on the Board of Directors of the Springfield Savings Society.

He was well known for many kind and thoughtful deeds for those in need. The following article from *The Springfield Daily News* dated December 23, 1910 states it well:

KLEEMAN PROVIDES FOOT WEAR

A number of boys suffering for lack of good shoes were made happy by Max Kleeman yesterday, who ordered Superintendent Parsons of the Detention Home, to take eight boys from the street, who needed shoes, and have them fitted. The boys were easily found and taken to O'Neil's shoe store, where Mr. O'Neil added six pairs, which will be distributed by Mr. Parsons today.

He was also remembered for his charitable contributions to the Dominican Sisters of the Poor, and to crippled and underprivileged children. He was a member of the Knights of Pythias, Benevolent and Protective Order of Elks, H.S. Kissell Lodge No. 676, Free and Accepted masons. They were both members of the South Fountain Avenue Jewish Temple.

After a short illness, Max Kleeman died on February 3, 1940, aged 72 years, 7 months and 27 days. After his death, his wife, Bess, moved to Edgewater Beach near Chicago where she died January 23, 1949. They are together in Section G, Lot 256.

Arthur W. Aleshire
Myrtle J. Marsh
(February 15, 1900 - March, 1940)
(October 9, 1898 - September, 1982)

Arthur Aleshire was born to James W. and Ada Painter in Luray, Page County, Virginia on February 15, 1900. When he was 11 years old, the family (which included several siblings) moved to Tremont City in Clark County, Ohio. Aleshire's early years were spent attending school and helping with farm chores. On May 11, 1922, he wed Myrtle J. Marsh, daughter of South Vienna residents Newton and Jessie Farish Marsh. The couple had one son.

Aleshire continued to labor on his 80-acre farm and in the successful operation of a local filling station until he was seriously injured in a truck accident. Tragically, his back was broken, resulting in paralysis. Following months of hospitalization Arthur resumed his life, as much as possible, with the aid of a wheelchair and close friend Forest Chamberlain, his constant companion.

He became politically inspired and ran for office in 1935, despite his physical challenges. Perhaps due in part to his courage and strong-willed example, Aleshire was elected to Congress on November 3, 1936, representing Ohio's 7th district. During his two years in office, labor and farm issues were his main focus. A member of the

House Census Committee, he also served on the House Committee on Election of President, Vice President and Representatives in Congress.

Upon his defeat in the ensuing election, Aleshire returned home and continued his political involvement. As the titular head of his party, Aleshire frequently hosted political conferences among districts throughout Ohio. His chronic poor health and accompanying collateral illnesses worsened near the end of 1939. He lingered several months in critical condition before dying in March of 1940 at the age of 40. He's buried alongside family members in Section U of Ferncliff Cemetery.

Aleshire's *Springfield News* obituary, dated Tuesday, March 12, 1940, quoted close friend and Postmaster W.A. Circle as saying, "Mr. Aleshire possessed good common sense and used it. He was a kind and considerate gentleman whose passing is a distinct loss to this community."

Harry A. Toulmin Sr.
Rosamond Evans
(1859 - March 25, 1942)
(1868 - October 5, 1947)

Harry Aubrey Toulmin was born in Toulminville, near Mobile, Alabama in 1859. His father, Morton Toulmin, was a manufacturer and cotton broker prior to the Civil War. While still a young boy, Toulmin moved with his family to New Orleans, where he received elementary schooling. Later, the Toulmins traveled to Washington, D.C., where Harry joined the law firm of Alexander and Mason while studying law at National University. He graduated in 1882 and was admitted to the bar one year later.

Toulmin wed Rosamond Evans, daughter of Dr. Warwick and Mary Mason (Washington) Evans of Washington, D.C., in April of 1887. Two children were born: Warwick Morton, who died August 6, 1889 at 16 months of age, and Harry Aubrey Jr.

By 1890 Harry Sr. was arguing cases before the United States Supreme Court and various U.S. circuit courts, as well as the Supreme Court of Cuba. He maintained his Washington, D.C. office until being offered a retainer from Springfield manufacturers to settle in Springfield, where he established a patent law firm and laid the foundation for a practice that extended throughout the United States and abroad. Toulmin arrived with his mother and his wife, Rosamond, and worked from an office inside the Bushnell Building.

In 1904 Orville and Wilbur Wright met with Toulmin for patent advice, laying the groundwork for an attorney-client relationship that weathered nine years of patent applications and patent-protection lawsuits.

Locally, Harry Toulmin Sr. was active with

the City Hospital and Snyder Park. He moved his law offices to Dayton, Ohio in 1911 to work more closely with the Wright Brothers, and was eventually joined by his son in the new firm of Toulmin & Toulmin.

Harry Toulmin Sr. died in Dayton, Ohio on March 25, 1942. His wife, Rosamond, followed on October 5, 1947, and their son, Colonel Harry A. Toulmin Jr., on March 28, 1965. His parents are also buried with the family in Ferncliff Cemetery, Section C, Lot 33.

David Orrin Steinberger
(March 25, 1857 - December 12, 1945)

David Steinberger was born on March 25, 1857 in Champaign County, Ohio, one of six children born to George and Barbara Elizabeth Funk Steinberger. He attended a nearby common school at an early age, and during the 1873-1874 school year enrolled in the Wittenberg Preparatory Department. By the following year he was in the Sub Freshman Class. Steinberger later studied at the National Design and Art League Schools in New York City, and in 1894, returned to Wittenberg College to teach art.

During the Spanish-American War (1898-1899) Steinberger drew the "Accoloid," otherwise known as an allegorical picture of Cuba, and donated the drawing to the American Red Cross.

Around 1900 he was stricken with tuberculosis. Believing a cure could be found if he moved west, Steinberger traveled to Colorado for a time, but returned to his father's farm having made the decision to live the remainder of his life outdoors. A tree house he dubbed "Camp-A-Loft" was constructed for him on the farm. Steinberger's health improved considerably with outdoor living—so much so that he believed it to be a cure and never again lived his life surrounded by four walls.

Steinberger became known as "the man who lived in a tree" and (more familiarly throughout Ohio) as "the Hermit of Mad River."

He died on December 12, 1945 in his native Champaign County at the age of 88. This talented and eclectic personality is buried in Section H, Lot 119 of Ferncliff Cemetery.

Harry S. Kissell
Olive Troupe
(September 25, 1875 - February 14, 1946)
(1878 - July 29, 1962)

The Nicholas Kissell family emigrated to America during the early 1700s as members of a Moravian Mission, traveling from their home in Wurtenburg, Germany to settle first in Pennsylvania and, later, in Maryland. Emmanuel M. Kissell, grandfather of Harry S. Kissell, was born in 1822 near Chambersburg, Pennsylvania and received most of his schooling in Quaker educational institutions.

Shortly after the family moved to Ohio, Emmanuel established a grocery business on the corner of High Street and Fountain Avenue. He eventually discontinued the business and delved into inventing and manufacturing. Inside his small plant on West Main Street, near Wittenberg Avenue, he developed what was known as the Kissell plow and continued his agricultural machinery improvements for a number of years. He patented a glass feeder for grain drills and fertilizers later manufactured by the P.P. Mast Company.

In 1872 Emmanuel Kissell entered the real-estate field in partnership with his father, C.D. Kissell. After C.D.'s death, he operated the business alone until his retirement, at which time his son and grandson, C.B. and Harry, became its successors. Emmanuel was an active member of the Springfield Order of Odd Fellows #33, and at the time of his death in 1910, held the distinction of being its oldest member. Notably, he also achieved status as one of Ohio's oldest judges, serving at age 87 as Democratic election judge in Precinct E of the Fourth Ward, a position he'd held for several years. Emmanuel Kissell died on May 27, 1910.

Cyrus Broadwell (C.B.), son of Emmanuel and Abigail Kissell, was born on February 11, 1849 in Leitersburg, Maryland. In 1851, the family journeyed by stagecoach to Ohio, where Cyrus married Lucretia Caroline McEwen on December 25, 1872 in Shelbyville. Shortly after the wedding the new couple moved to Springfield, where Cyrus entered the real estate business with his father and rose to prominence due to his energy and personal integrity. He remained active in the community as a member of the Second Presbyterian Church, the Anthony Lodge, F. & A.M., and with Palestine Commandery, Knights Templar #33. Following a short illness in the fall of 1903, Cyrus died in his South Fountain Avenue home on October 29 at the age of 54. His wife followed on March 5, 1930 at the age of 81. Daughter Blanche married Ralph Busbey. Son Harry Seaman was born on September 25, 1875.

Harry S. Kissell obtained a solid education, graduating from Wittenberg College in 1896. After studying law for two years and working as a newspaper reporter he entered the family-owned Kissell Real Estate business with his father. He married Olive Troupe, daughter of Theodore and Mary C. Winger Troupe, in Springfield on October 17, 1901. Two children were born: Roger Troupe and Mary Lucretia.

In 1907 Harry met with a small group of real estate dealers in Louisville, Kentucky and helped organize the National Association of Real Estate Boards (in addition to the Ohio and Springfield Board of Realtors). He served on various National Association committees and on its board of directors, and was also selected to several terms as president and vice president.

Mr. Kissell was one of the original organizers of the American Trust and Savings, which merged in 1928 with First National Bank, and served continuously on the directorate of both financial institutions until his death. He directed the Clark County Chapter of the American Red Cross and served on the Ferncliff Cemetery Association, was a Wittenberg College executive committee member, and a founder and member of Ridgewood School.

The Springfield Community Fund became a successful operation under Kissell's leadership. He served as its first chairman in 1923 and held the position for many years. He helped establish the Springfield Rotary Club, for a time served as its president, took an active interest in the organization's statewide charitable efforts on behalf of crippled children, and played a key fundraising role as a member of the Ohio Society for Crippled Children.

His other interests included membership in Beta Theta Pi fraternity, the Springfield Polo Club, the Springfield Country Club, the Springfield Chamber of Commerce, and the Sons of the American Revolution (ancestor George Kissell fought in that war).

In 1910 at the age of 35, Kissell was elected Most Worshipful Master of Masons in Ohio. By 1925, he headed a $500,000 effort to construct a Masonic Temple in Springfield, and during his lifetime, received the rare honor of having a Masonic Lodge bear his name.

Kissell and his wife were active members of Covenant Presbyterian Church, with both serving in various capacities. Harry, in particular, served as deacon, elder, trustee, and president of the board. He continued his civic and humanitarian work into his 70th year, until stricken by a fatal heart attack during a Cincinnati business meeting on February 14, 1946. The famed businessman is interred in Ferncliff Cemetery alongside his wife in Section O, Lot 56.

Dr. Bennetta D. Titlow
(January 24, 1868 - March 7, 1946)

Dr. Titlow was born in Lancaster County, Pennsylvania on January 24, 1868, one of three children of Jerome and Hettie Davis Titlow. His siblings, Harriet Woodfin Titlow and John Titlow, were born in Hampton County, Virginia.

Details of Jerome's early years are sparse, with the exception of his Civil War service in Company K, 3rd Pennsylvania Heavy Artillery. He was the captain in charge at Ft. Monroe, Virginia on May 23, 1865—the day that Jefferson Davis, former president of the Confederacy, was accused of inciting the assassination of President Abraham Lincoln and ordered into arms. Captain Titlow personally supervised Davis' shackling.

Dr. Chester D. Bradley, Civil War museum curator at Ft. Monroe, authored an article entitled "Jefferson Davis in Prison" that was published in the Spring, 1958 edition of *Manuscripts* magazine. The piece offered post-war information about Captain Titlow. Likewise, the *Springfield Daily News* of August 30, 1959 provided facts offered by Mary Miller of Warder Public Library.

Sometime during the 1880s the Titlow family moved to South Dakota and eventually settled in St. Paul, Minnesota. Captain Titlow died on October 13, 1912 in the Minnesota Soldiers' Home while his family was residing in Springfield, Ohio. His wife, Hettie, died on January 6, 1926, after a prolonged illness. Before her health failed, she'd been active in Christ Episcopal Church and the Woman's Club.

Dr. Bennetta Titlow graduated from the Women's Medical College of Philadelphia, Pennsylvania (Class of 1891) and first practiced medicine in Springfield in 1892. She was instrumental in creating the "Baby Health Camp" launched before World War I and, at the time, located on East Main Street near Standpipe Park. The project offered public health care for infants and children and was afforded general support during its existence.

Dr. Titlow joined the Red Cross during the war and served one year with Medical Corps in France. An active member of the Clark County Medical Society in 1929, she authored an outstanding paper on "Spontaneous Pneumothorax." Her affiliations included the American Medical Association, Ohio State Medical Association, and the Altrusa Club.

Dr. Titlow died on March 7, 1946, having

practiced medicine in Springfield for 47 years. Her sister, Harriett, who retired as supervisor of art in Jersey City Schools in 1934, preceded her in death on December 27, 1943. John, a printer and former editor of *The Sun*, died on January 4, 1954. They are buried together in Section Q, Lot 171 of Ferncliff Cemetery.

Sully Jaymes
(March 13, 1880 - January 20, 1950)

Sully Jaymes, son of Hamilton and Ella Anderson Jaymes, was born in Campbell County, Virginia. He graduated in 1896 from English High School in Boston, Massachusetts, and in 1898 from Boston University. He attended Boston University Law School before graduating from the University of Michigan in 1901, an L.L.B. degree in hand.

Jaymes established a Columbus, Ohio-based law practice and in 1903 arrived in Springfield to continue his career. A man of distinction, he served as director of the Charles Funeral Supply Company of Springfield, president of James Cottman Real Estate, and was a former Wilberforce University trustee. For six years he held the title of grand attorney for the Grand Lodge, Knights of Pythias of the State of Ohio and later became Grand Chancellor of Ohio in 1934—a position he held until his death. His memberships included Mystery Lodge #45, K of P, Springfield, the Clark County Bar Association, Greater Springfield Association and North Street A.M.E. Church. Jaymes also held a seat among the inaugural board of directors for the Center Street YMCA.

Jaymes became ill in December of 1949 and died on in January of the following year, aged 69 years. His wife, Anna B., died September 14, 1960, aged 74 years. They are buried together in Ferncliff Cemetery, Section Z, Lot 66.

James Garfield Stewart
(November 17, 1880 - April 3, 1959)

The eighth child of Colonel James E. and Mary E. Durbin Stewart, James Garfield Stewart was born in Springfield, Ohio on November 17, 1880 and received his early education in Springfield schools. He later attended Kenyon College in Gambier, Ohio (Class of 1902) and Cincinnati Law School (Class of 1905). Stewart practiced law in his hometown until 1908, then moved to Cincinnati where he argued cases before state and federal courts and, later, the United States Supreme Court.

Stewart married Harriet L. Potter, a Jackson, Michigan native, on September 7, 1910 and the couple had three children.

After years as a leading Cincinnati trial lawyer,

Stewart entered politics and served several years as an elected city councilman. He served as Cincinnati mayor from 1938-1947 and as State Supreme Court Justice from 1947-1959.

Judge Stewart was a member of the Cincinnati, Ohio State and American Bar Associations, Alpha Delta Phi, Phi Delta Kappa and Phi Delta Phi fraternities, the Cincinnati Club, Cincinnati Country Club, the University Clubs of both Cincinnati and Columbus, and the Cuyler Press Club. He was a 33rd-degree Mason and both a Scottish Rite and York Rite member.

Stewart suffered a fatal heart attack while en route to a speaking engagement and died on April 3, 1959 in Louisville, Kentucky. His body was returned to Springfield and interred with family members in Section D of Ferncliff Cemetery.

Gus F. Sun Sr.
Nellie Alfrey
(October 7, 1868 - October 1, 1959)
(1883 - February 13, 1953)

Gus Sun was born Gustave Ferdinand Klotz on October 7, 1868 in Toledo, Ohio, the fifth child of Johan and Louise Klotz, hotel owners who catered to theatrical groups. Given his influential surroundings, it was perhaps inevitable that Gus, along with brothers John, Peter and George, fell in love with the entertainment world.

At the age of 15, Gus traveled to New York and found work in a poolroom, setting up balls and performing his juggling act. He returned home, however, after George was injured in a circus act.

Enduring considerable pressure from his father to "learn a useful trade," Gus the eclectic, free spirit went to work as an apprentice mechanic in a sewing factory—a seemingly awkward fit. Predictably, he continued his juggling act at night. His first break came when he was booked in a minstrel show as Herr Gus Klotz for $1 a performance. He later joined a medicine show, crisscrossing the country to sell Indian herb medicines while playing the role of the doctor, all the while winning over audiences with his juggling ability.

Gus spent two more years as a vaudevillian juggler, appearing in circuses and variety shows, before he and his brothers organized the Sun Brothers United Shows and Trained Animal Exhibition. Soon known as the "Largest Wagon Show on Earth," they toured the country successfully for many years.

While touring California in 1904, Gus became interested in motion pictures. Upon his return to Ohio he moved to Springfield and opened a small theater known as the Old Orpheum, combining vaudeville and motion pictures and showing three to five acts daily. The venture proved such a phenomenon that he quickly opened a second theater in 1908—the New Sun Theater, later renamed the Ohio Theater. The Alhambra Theater followed in 1913, the Fairbanks in 1917, and the Regent Theater in 1920.

Sun found himself overwhelmed by requests from other theater owners in search of talent, and by 1908 he was booking acts into 50 theaters. His hand-picked actors included Walter Huston, George Burns and Gracie Allen, Fanny Brice, Bob Hope, Fibber McGee and Molly, and Al Jolson—his eye for talent helped make stars of them all. By 1920, Sun's business deals with other vaudeville circuit managers increased the number of Sun-affiliated theaters to 275.

With the invention of "talkies" Sun was forced to close a few of his live-performance theaters and began booking acts for carnivals and fairs. The August 11, 1926 issue of *Variety* was devoted entirely to the success of Gus Sun, with his worth estimated at several million, staggering for the times. By doing what he loved best, despite his father's early displeasure, Gus Sun became an entertainment institution.

Gus Sun Jr. took over his father's business in 1926, while Sr. made the most of his later years by traveling to far and distant places such as Europe and Africa. He fell and injured his hip on September 10, 1959 and was hospitalized. Pneumonia set in and, three weeks later, on October 1, 1959, the renowned showman died. He is buried beside his wife, Nellie, in Section O, Lot 154 of Ferncliff Cemetery.

Davey Moore
(November 1, 1933 - March 25, 1963)

Born in 1933 in Lexington, Kentucky to Reverend Howard Thomas and Jesse Moore, David S. "Davey" Moore moved with his family to Springfield when he was a small child.

As a youngster Davey showed great promise in the boxing world when he and his friend, Higgins, performed in the ring during intermissions of Springfield Golden Gloves tournaments at Memorial Hall. With each appearance he matured and honed his skills.

Moore was 20 years old when he made his boxing debut on May 11, 1953, beating Willie Reece in a six-round decision. Through 1956 he won 22 of his first 27 fights (with one no-contest). He joined the U.S. Olympic team in Helsinki in 1952.

Married at the time and supporting a growing family, Moore became discouraged with boxing's low pay and entertained notions of quitting. Manager Willie Ketchum persuaded him to keep fighting, however, and Moore's boxing career eventually took the pair to France, Spain, Italy,

Finland, Venezuela and countless other countries. Moore strung together 18 consecutive wins before before suffering a TKO to Carlos Hernandez on March 17, 1960. Moore's highly publicized, first-round knockout of Bob Gassey three days earlier had claimed all but two of Gassey's teeth.

It was during that win streak and accompanying notoriety that the fleet-footed Springfielder dubbed "Little Giant" earned a much-anticipated shot at World Featherweight champion Hogan "Kid" Bassey—a March 18, 1959 bout that Moore won with a 13th-round knockout. A year later, Moore claimed the much-anticipated rematch with a decisive, 11th-round blow. He successfully defended his World Featherweight Title until a fatal, nationally televised bout with undefeated, No. 1 contender Ultiminio "Sugar" Ramos on March 21, 1963 in Los Angeles, when he collapsed in the 10th round. News reports from inside Moore's dressing room described the famed champion as not appearing seriously ill, sporting only a bloodshot left eye and talking for some 40 minutes after the fight, saying, "Sure, I want to fight him (Ramos) again. And I'll get the title back."

Sparring partner Ronnie Wilson later recalled Moore placing his hand to his head and saying, "Oh, my head aches" before the 29-year-old boxing great fell unconscious. He never recovered and died on March 25.

Moore's unexpected death sent shock waves around the world, prompting legendary folk singer Bob Dylan to compose a righteous song in his honor, "Who Killed Davey Moore?" Preserved for the ages on *The Other Side of the Mirror*, a poignant DVD collection of Dylan's Newport Folk Festival performances from 1963 to 1965, the tune was penned shortly after Moore's death.

Numerous books and articles have also been published about the deadly Moore-Ramos fight. Moore's final statistics vary among professional sports sources, with some reporting as many as 59 wins and 30 knockouts. According to the *Springfield Daily News*, Moore's lifetime professional record was 56 wins, 6 losses, and one draw, with 28 knockouts. A brief, biographical sign today greets visitors of Davey Moore Park.

Ralph Davis
Edna M. Brougher
(August 11, 1898 - June 4, 1978)
(June 8, 1897 - February 28, 1984)

Ralph Davis, the son of John D. and Mary Catherine Spriggs, was born in Wellston, Ohio. He eventually married Edna Brougher, the daughter of Andrew and Laura Link Brougher, born on June 8, 1897 in Tradersille, Ohio. Davis founded the Eagle Tool & Machine Company in 1939, providing employment to many throughout Clark County. He and his wife, Edna, who also served as company co-owner, built the company into the success that it remains today.

Both were members of the Central Methodist Church, while Ralph remained active in the American Society of Tool Engineers, Anthony Lodge #455 F. & A.M., Springfield Chapter #48 RAM, Springfield Council #17, R. & S.M., Palestine Commandery #33 Knights Templar, Antioch Temple Shrine, and the Springfield Shrine Club. Edna, meanwhile, contributed her spare time to the Neal Chapter, Order of the Eastern Star.

Ralph Davis died on June 4, 1978 following a six-week illness. Edna followed on February 28, 1984. They are interred in Section X, Lot 34 of Ferncliff. A large monolithic-style sarcophagus adorns their grave. The monument's epitaph features a pair of eagles and the inscription "Founder of Eagle Tool & Machine Company."

Robert C. Henry, second from right at the White House with President Richard Nixon.

Robert C. Henry
(July 16, 1921- September 8, 1981)

Robert Clayton Henry was born in Springfield, Ohio on July 16, 1921, one of twelve children born to Guy L. And Nellie Reed Henry. His early years were spent attending the local schools and Wittenberg University. With a desire to enter the mortuary business he went to the Cleveland College of Morturary Science in Cleveland, Ohio for his degree in mortuary studies. By 1951 Robert Henry achieved his goal when he founded the Robert C. Henry Funeral Home. He was married to Betty Jane Scott Henry and they had three children.

Apart from his mortuary business, Robert was active in charitable causes and served as head of many charity drives and civic organizations in Clark County. In 1961 he entered the local political field, and in his first campaign was elected to the Springfield City Commission. Four years later the commission appointed him as the city's mayor in 1966. Upon his installation as mayor in 1966, Robert Henry was nationally recognized as the first black mayor of a city the size of Springfield, Ohio. In 1968, he refused to run for re-election but remained on the commission. That same year he was awarded an honorary Doctor of Humane Letters degree from Central State.

After completion of his term as mayor, Robert Henry was selected as a member of a fact finding commission to Vietnam by order of then-president, Lyndon Johnson. He later returned in 1970 under Richard M. Nixon to inspect non-military activities. In 1972 he was the Republican Party nominee for the 60th District seat in the Ohio House of Representatives, then lost in the general election.

In 1976 the International Harvester Company bestowed special recognition on Robert Henry for the major role he played in the Company's negotiations with the city on water line extension and other details needed for the new industrial complex for the new IH Truck Assembly Plant.

He continued to be active in many charitable activities as well as the operation of the mortuary business with his family until illness overcame him, and he died of cancer on September 8, 1981. He is interred in Ferncliff mausoleum.

He was honored in many ways including a parade and banquet on Robert C. Henry Day during Black History Month, the downtown fountain dedicated to his memory and a retirement home that bears his name.

Paul U. Deer
Elizabeth Kelsey
(May 26, 1897 - January 4, 1984)
(September 9, 1898 - June 8, 1995)

The son of George W. and Ida Litsey Deer, Paul Deer was born on May 26, 1897 at Deer's Mill, Indiana. He attended Crawfordsville High School and Wabash College, and became a member of the American Expeditionary Force during World War I.

Deer and his father formed George W. Deer & Son Corporation, an Indiana-based gasoline and motor oil distribution company, in 1922 and later sold the entity to Cities Service Oil Company in 1929.

He married Crawfordsville native Elizabeth Kelsey, born September 9, 1898 to Hugh and Amelia (Harrison) Kelsey. After the couple moved to Springfield in 1932 Deer launched Bonded Oil Company, opening his first station one year later in Urbana, Ohio. Soon after, three stations were added in Springfield. Bonded Oil merged with Marathon Oil Company in 1975. At the time of the merger there were 208 Bonded gasoline stations.

In 1967 oil industry notables elected Deer to the Oil Hall of Fame. Deer was honored in 1971 when the Ohio Petroleum Marketers Association recognized his fifty years in the oil industry. He was among the founders and initial board members of the Independent Oil Men of America, Branded Petroleum Marketers of America, and the Society of Independent Gasoline Marketers of America who honored him in 1977 as "Dean of Private Brand Marketers in the United States."

Locally, Mr. and Mrs. Deer operated the 1,300-acre Paul Deer Farm, then just east of Springfield, and they were nationally recognized among the leading breeders of registered Polled Hereford cattle. The Deers were active in numerous community organizations and gave generously to such institutions as Wittenberg University, the United Way, Clark State Community College and Community Hospital. The Paul and Elizabeth K. Deer Exposition Hall at the Heritage Center of Clark County memorializes the couple's many contributions to the community.

Paul Deer died on January 4, 1984 at the age of 86. Elizabeth died on June 8, 1995 at the age of 96. They are buried in Ferncliff Cemetery, Section O, Lot 161.

Merlin Gladstone Robertson
Elenore Wachs

(April 26, 1896 - September 26, 1985)
(December 8, 1898 - November 5, 1994)

The son of William Frederick and Westanna O'Neill (Brown) Robertson, Merlin was born in Cincinnati, Ohio on April 26, 1896. His father, a successful businessman, operated a warehousing and distribution center that specialized in wire products for the hardware trade. As his business grew he explored new avenues, purchasing the Elwood Myers Company in Springfield, a producer of metal and leather advertising novelties. The business was soon renamed Robertson Steel & Iron Company, Elwood Myers Company division.

William Robertson also acquired a business based in Madison, Indiana that manufactured small nails and brads, as well as a fence manufacturing company in Cincinnati, Ohio. He later added other businesses to his manufacturing holdings and dabbled in real estate, at which he excelled.

Although Robertson's main enterprises were in Springfield, he continued to live in Cincinnati. During the early 1930s his health declined and he became bedridden for the last few years of his life. He died in 1936, and Merlin assumed control of the family businesses. Merlin was not, at this time, associated with the Robertson Company, but was operating the Dayton Yellow Cab Company, which he'd founded years earlier.

Merlin Robertson and his family moved to Springfield and, in the aftermath of business consolidations, he became president of the Robertson Can Company, maintaining control until he retired in 1970 and moved to Tryon, North Carolina.

Merlin and his wife, Elenore, the daughter of William Carl and Sadie Glover Wachs, were quite generous over the years, contributing to such organizations as Covenant Presbyterian Church, Buckeye Trail Girl Scout Camp, the Springfield

Beautification Committee, Springfield Symphony, Springfield Art Museum and Clark Memorial Home.

Merlin died following a brief illness on September 26, 1985 in Tryon, leaving behind his wife and four children. His body was returned to Springfield for burial in Ferncliff Cemetery.

Elenore, a Cincinnati native and graduate of the University of Cincinnati's College of Music, spent much of her life as a talented violinist and Springfield Symphony member. She was also a charter member of Clark Memorial Home. Elenore remained in North Carolina until her death on November 5, 1994. She is buried beside her husband in Ferncliff's Section O.

Margaret Evelyn Baker
(May 29, 1896 - January 10, 1986)

Margaret Evelyn Baker, only child of Scipio and Jessie Foos Baker, was born in Springfield, Ohio on May 29, 1896. Her parents hailed from wealthy, prominent families. Baker's mother was the daughter of John Foos, a man of prominence in banking and politics, and founder of the Foos Gas Engine Company that later became Mast, Foos & Company.

Baker attended Springfield Seminary and Springfield High School before graduating from Baldwin School at Bryn Mawr, Pennsylvania, then studied music in New York City. She appeared on concert stages throughout New York and other eastern cities for two seasons (1919-1921) under the stage name of Nancy Van Kirk, the "little Dutch girl in wooden shoes." She performed the *Children's Hour of Story and Song* at New York City's Lenox Little Theater, prompting the *New York Evening Post* to write, "Nancy Van Kirk has sympathy, richness of personality and artistry of expression, the qualities that children are quick to feel in her performances." Later roles as Mezzo Soprano in *An Hour of Chinese Song and Legend*, performed at New York's Princess Theater, were similarly considered outstanding.

Baker's musical career ended when her father died unexpectedly while on vacation. She returned to Springfield and became general manager of the Champion Company. Her mother passed in 1949, and Baker served as Champion's president until 1982, when she moved to San Diego, California. As a majority stockholder, she retained the title of "honorary president."

At one time, Margaret's holdings included a rubber plantation in Malaya as well as vested interest in the Springfield Equipment Company, the Tecumseh Building, Sonzonia Vault Company, Champion Company of Canada, Ltd., and a

Champion Company branch in California. At the same time, she was serving as president and treasurer of the Champion Company in Springfield.

Baker was highly active in local and national Republican politics. Her community activities included membership in the American Legion, Daughters of America, the board of directors of City Hospital Auxiliary, Women's Town Club, Business and Professional Women's Club and Altrusa International. She was the first president of Altrusa Club of Springfield, a pioneer women's service club whose members are prominent leaders in business, industry and professions. The Springfield Club, which Margaret helped to found, was the ninth club organized into Altrusa International.

From the age of five Baker traveled abroad and spent her fifth birthday in the Alps. By the time she became an adult she'd made 15 trips around the world, combining business interests with her hobby of travel-related, documentary film-making. Among her most well-known movies are *Formosa, Malay Peninsula, Thailand, Anna's Siam, Holland 1973, Primitive Africa*, and *Morocco and the Moors*. Many of her lectures and films were given in support of charity benefits.

Four years after moving to San Diego, Baker died there on January 10, 1986. Her body was returned to Springfield and interred with her parents in Section P, Lot 19 of Ferncliff Cemetery. She was the last survivor of her immediate family.

Richard O. Brinkman
(February 8, 1926 - April 7, 1986)

Born in Indianapolis, Indiana, the only child of Raymond J. and Irma Waltz Brinkman, Richard Brinkman spent his formative years devoting himself to his studies and obtaining a good education. He eventually earned a Bachelor's Degree from Wittenberg College and a Master's Degree from Kent State, and continued his studies at four other universities before graduating from the Harvard Institute for Educational Management.

Brinkman began his teaching career in Springfield and Clark County and for 33 years worked as a high school teacher, counselor, principal, local superintendent, public personnel services director, JVS superintendent and Clark Technical College president. Of his 30 years in administration, 21 were spent as founder and president of CTC.

He served in the United States Army during World War II and was awarded the Bronze Star for gallantry in action in Germany.

On August 22, 1948 he married Marguerite Conner, daughter of Charles and Maybell Pickett Conner, and the couple had two children: Douglas and Peggy.

Brinkman gradually involved himself in assorted community programs, serving as president of both the Springfield Area Chamber of Commerce and Rotary Club of Springfield, as director of Security National Bank and Trust Company, and as a park district commissioner. He was also a member of the Springfield Community Development Steering Committee, Clark County Facilities Study Group, Community Progress Council and Industrial Advisory Council of OIC.

Impressively, Brinkman found additional time to serve as president of both the Community Welfare Council and the United Way of Springfield and Clark County, and as chairman of the United Appeals Fund campaign. Honored for his 20 years of volunteer service with the United Way, Brinkman was named to the group's hall of fame.

Countless awards were bestowed upon the noted humanitarian, including the Governor's Award, an Ohio Senate Citation, and a Wittenberg University Alumni Citation. Brinkman Hall at the James A. Rhodes Career Center was named in his honor in 1974. Likewise, Clark Technical College's Downtown Center was dubbed the Richard O. Brinkman Education Center in 1983. He was also a Rotary International Paul Harris Fellow.

Brinkman collapsed and died while attending a meeting of the Springfield Rotary Club on April 7, 1986. An entire community felt his loss. His generous contributions to Springfield and Clark County can only be measured in time.

Appropriately, Brinkman's final wishes were fulfilled when he was laid to rest among fellow WWII soldiers in Ferncliff Cemetery.

Bradley Kincaid
Irma A. Forman
(July 13, 1895 - September 23, 1989)
(1897 - 1995)

Born in Point Level, Kentucky to parents William and Elizabeth Kincaid, Bradley Kincaid's initial interest was music. He played his first guitar as a young child growing up near the Cumberland Mountains, quickly parlaying his talent into radio stardom. His radio days began in 1926, when he performed on the *National Barn Dance* show on WLS (AM) in Chicago. Public adoration soon transformed "Bradley Kincaid and His Hound Dog Guitar" into an overnight sensation.

In the spirit of later homegrown country pioneers like Dolly Parton and Loretta Lynn, Kincaid's popular ballads paid poignant tribute to his Appalachian mountain heritage. He cut more than 150 songs during his career, including *Give My Love to Nell*, *Blue Tail Fly* and his ultra-memorable rendition of *Barbara Allen*, the latter arranged and recorded by Parton on her 1994 live CD, *Heartsongs,* and forever preserved in the Smithsonian Collection of Recordings of American Popular Song. Now one of America's most oft-collected and reinterpreted ballads, it was Bradley Kincaid who made the song legendary.

Kincaid, who famously dubbed friend Lewis Marshall Jones "Grandpa Jones," married Irma Forman and the couple had four known children. He arrived in Springfield, Ohio in 1950 from Nashville, Tennessee, where he'd spent five years performing at the historic Grand Old Opry. He visited in order to purchase a radio station, which he managed until it was sold. A later purchase, Morrelli's Music Store, was renamed Kincaid's Music & Audio.

In 1981, Loyal Jones, director of Appalachian studies at Berea College, authored a book entitled *Radio's Kentucky Mountain Boy, Bradley Kincaid*, which documented the singer's career and business dealings in Springfield.

Although he preferred to refer to himself as a folk singer, Kincaid was also a member of the prestigious Country Music Association and the Nashville Songwriters Foundation. So lasting and influential is his body of work that he was inducted into the NSF's hall of fame in 1971, joining such musical giants as: Gene Autry, Chuck Berry, Jimmy Buffett, Bob Dylan, Don & Phil Everly, Woody Guthrie, Merle Haggard, Buddy Holly, Loretta Lynn, Bill Monroe, Roy Orbison, Buck Owens, Dolly Parton, Carl Perkins, Marty Robbins, Ernest Tubb, Conway Twitty, and Hank Williams Sr..

Kincaid was twice nominated for entry into the County Music Association's Hall of Fame in 1980 and 1988, but lost to entertainment giants Johnny Cash and Roy Rogers. According to the Nashville Songwriters Foundation, son James said of his father's CMA Hall of Fame nominations, "Bradley Kincaid was twice nominated to the Country Music Hall of Fame, which was nice in that he had been retired for almost 40 years." A prolific composer, Kincaid's 1928 songbook *My Favorite Mountain Ballads* sold more than 100,000 copies.

(William) Bradley Kincaid died in Springfield on September 23, 1989 at the age of 94, having influenced generations of singer-songwriters to follow. Services were held inside Covenant Presbyterian Church, of which he was a member. He is interred in Ferncliff Cemetery, Section 28, Lot 44. Inscribed upon his monument is the following, fitting tribute:

The Kentucky Mountain Boy
Pioneer Radio Entertainer
Whose mountain ballads filled the air touching the hearts of all who listened.

Carleton F. Davidson
(February 3, 1894 - November 25, 1993)

Carleton Davidson was born in Peru, Indiana to Frederick and Elva King Davidson on February 3, 1894. He later attended the University of Cincinnati before enlisting in the army during World War I and serving as a first lieutenant. Upon his arrival in Springfield, Davidson founded Dayton Chevrolet Company, a successful dealership venture that thrived for many years.

An active citizen, he concerned himself with local affairs as both a Springfield city commissioner and, later, mayor (1954). He served on boards of directors for Community Hospital, the YMCA and M&M Savings and Loan, was a life member of the Lions Club, and also joined St. Andrews Lodge #619.

A philanthropist dedicated to many Springfield and Clark County causes, Davidson established the Carleton & Ruth Davidson Trust Fund to help finance Carleton Davidson Baseball Stadium, which opened in April of 2004. He also contributed richly to the Davidson Interpretive Center located near George Rogers Clark Park. The beautiful center opened in November of 1999 and today tells the interpretive story of the Battle of Pickuwe and Clark's frontier effort to drive out native Shawnee Indians.

Carleton Davidson died in his residence on November 25, 1993, three months shy of 100. He left behind a philanthropic legacy to be enjoyed by generations to come. He and his wife, Ruth, who died in 1990, are interred in Ferncliff Cemetery's Section O, along with Davidson's mother, twin sister, Alice, and brother, Joseph.

Harry Morris Turner
Violet Zimmerman
(April 22, 1903-January 10, 2000)
(October 7, 1904-February 3, 1985)

Harry Turner was born in Atlanta Georgia to Charles E. and Elizabeth Everett Turner on April 22, 1903. In the 1920s his father's failing eyesight brought the family back to their roots in Ohio where Harry met and married Violet Zimmerman, daughter of Frederick and Goldie Patton Zimmerman and started raising a family. Violet had received a good education and was a dedicated teacher in both the Springfield City Schools and

Ridgewood School for a number of years.

As a young man, Harry Turner worked at several jobs then entered the insurance business in 1924, working for others until the collapse of the economy. He then started the Turner Insurance Agency in 1932, and over the years merged with several other local agencies. In 1949, Mr. Turner together with Jack and Bob Schiff, and Chet Field founded Cincinnati Insurance Company, which later became Cincinnati Financial Corporation. In 1961 the Turner Insurance Agency merged with Fred Wallace Agency, forming Wallace & Turner Incorporated.

Harry was instrumental in the organization of the first Community Chest in Springfield, now the United Way, and active in the Springfield Area Chamber of Commerce. He was an active member of Anthony Lodge 455 Free and Accepted Masons, Springfield Chapter 48 Royal Arch Masons, Springfield Council 17 Royal and Select Masons and Palestine Commandery 33 Knights Templar. Harry and Violet were active members of Central Christian Church, where Harry served as a trustee.

Harry and Violet Turner were contributors to many charitable causes in Springfield including Central Christian Church, Clark State Community College, the Heritage Center of Clark County, Springfield Symphony Orchestra, Springfield Museum of Art, and many student scholarship programs. Harry and his wife were also quiet contributors to many other charities and individuals.

After several years of failing health, Violet died February 3, 1985. Harry Turner continued with his philanthropies until his death following a short illness on January 10, 2000 in Sunrise Florida. His last humanitarian gesture was to leave a large portion of his estate to establish the Harry M. and Violet Turner Charitable Trust in order to give back to the community that he loved. He lies in rest next to his wife in Ferncliff Cemetery, Section X.

Brooks Lawrence
(January 31, 1925 - April 27, 2000)

The fifth of six children born to Wilbur and Patsie Walker Lawrence, Brooks Lawrence quickly joined the ranks of Springfield's baseball greats.

The multi-sport standout attended Western School and Keifer Junior High before graduating from Springfield South in 1943, having excelled in track, basketball and football at Keifer. The gifted athlete later etched his name into local sports lore as a first-string quarterback for the Wildcats.

Lawrence's budding sports stardom was

temporarily disrupted by his World War II army service in Guam, but he returned home having earned a Bronze Star. He secured post-war work at the National Supply Company and in 1947 enrolled at Miami University. Within a year he was playing on the school's collegiate baseball team.

Lawrence soon requested and received a tryout with the Dayton Indians ... and a professional baseball career took root. He spent time in the minor leagues until his major-league debut in 1954, playing for the St. Louis Cardinals until a trade to Cincinnati was completed a year later. The hard-throwing pitcher won his first 13 games with the Reds, was voted to the National League All-Star Team, and ended the year as the club's top hurler with 19 victories.

Lawrence married Larcenia "Dolly" Winston and the couple had two children: Anthony and Patsy.

A Cincinnati and Ohio Baseball hall of famer, Lawrence eventually retired from the sport and worked locally for 10 years at International Harvester before securing a front-office job with the Reds. Notably, he oversaw scouting, minor-league player development, radio and television work and season-ticket sales as the club's first African American hire.

Brooks Lawrence died of cancer on April 27, 2000 at the age of 75. He is interred in Ferncliff Cemetery's WWII annex.

Michael H. Chakeres
(1912-2002)

The late Michael H. Chakeres had an exceptionally praiseworthy career in the theatrical world throughout his lifetime. He was born in Dayton, Ohio, the son of Harry and Vasiliki Chakeres, then moved to Springfield, and graduated from Springfield High School and Wittenberg University.

His career started as a young lad serving popcorn to the patrons in his father Harry's theaters, later as assistant manager. During World War II he served three years as Chief Warrant Officer in the United States Air Force. When he returned home he continued the family business until his father's death in 1955 when he took control, became President and Chairman of the Board of Chakeres Theatres and Lobby Shoppes, Inc.

Michael and his wife, Pauline Dombalis Chakeres and children Valerie, Philip and Harry N. Chakeres, were devout members of the Assumption Greek Orthodox Church in Springfield where Michael was president of the Council. He was also a member of the Greek Orthodox Church of Dayton, Columbus, Ohio and Palm Beach Florida., the Leadership 100 of the Greek Orthodox Church of the Americas, and The American Hellenic Education Progressive Association. In 1992 he was made "Archon" of the church, the highest honor bestowed upon a layman, and recipient of the "2000 Ellis Island Medal of Honor" from the National Association of Coalition of Organizations Foundation, Inc.

He served on the Board of Directors of Springfield Key Bank, Community Hospital, Independent Endowment Fund, Wittenberg University, the Springfield Foundation, and many other organizations. He devoted his life to many philanthropic causes and strove to make a difference in local and national charities.

He was an active member of the Rotary Club, University Club, Van Dyke Club, Scottish Rite Valley of Dayton, 32nd degree Mason of the H. S. Kissell Lodge 674 Free and Accepted Masons, and continued his charitable activities until his death at the age of 90 on December 7, 2002.

He lies in peace in Ferncliff Cemetery, Section X.

Harry Thomas Lagos
Eugenia Papas Lagos
October 1, 1912--May 12, 2002
December 23, 1917- November 26, 2005

Athanasios (Tom) Lagos, a young man from Greece, immigrated to the United States in the late 1800s. Sometime in the 1890s he settled in Springfield, Ohio. By 1899 he and his brother had already established themselves with the Lagos Brothers Ice Cream and Candy store located on the northwest corner of Fountain and High and another store on West Main Street until his return to Greece in 1912.

Harry Thomas, son of Athanasios (Tom) and Garifalia (Lillian) Lagos was born October 1, 1912 in Geraki, Laconia, Greece, a small mountainous village in the Peloponnesian Peninsula. He and his siblings attended the village school in their primary years, with Harry completing the fourth grade. With his innate ability and determination to succeed in life, he pursued every opportunity to attain that goal. His motto "Columbus took a chance" was uppermost in his mind in 1931 when he arrived in the United States hoping to achieve the American Dream. He taught himself to speak English, obtained work and eventually opened his own business. He became a real estate investor and was in partnership in the Elite Café for 47 years.

During World War II, Harry saw service in the Army under Generals Omar Bradley and Mark Clark, attaining rank of Staff Sergeant. His battalion of Sherman tanks was a unit specializing in invasions and participated in the invasions of North Africa, Sicily, Italy and southern France.

On August 31, 1947 Harry married Eugenia Papas daughter of James (Demetrios) George Papas (Papathanassiou) and Lillian (Ligeri) Sarros Papas, both from villages in the Peloponnesian Peninsula in the southern part of Greece. Eugenia was born in Sidney, Ohio on December 23, 1917, attended the public schools and was valedictorian of the Sidney High School Class of 1935. She received full scholarships from several universities but declined them as she was not permitted to attend college by herself out of town. During and after high school, she worked in the family restaurant until her marriage. She, being an excellent pianist, founded a music group promoting classical music in Sidney, Ohio.

There were three children from this union, Thomas Harry, James Harry, and Litsa Lagos Press. Following their father's advice of studying hard and working hard contributed to their successful careers.

Both Harry and Eugenia were charter members of the Greek Orthodox Church Assumption of the Blessed Virgin Mary, with Harry having served on the Parish Council including Vice President, and Eugenia as Sunday School teacher for many years. They also contributed to many charitable causes.

After a brief illness Harry passed from this life on May 18, 2002. Eugenia followed him in death on November 26, 2005. They rest together in Ferncliff Cemetery, Section 32.

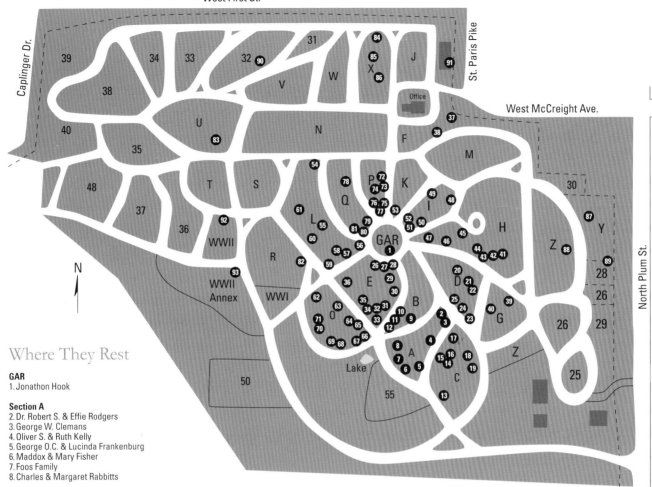

Where They Rest

GAR
1. Jonathon Hook

Section A
2. Dr. Robert S. & Effie Rodgers
3. George W. Clemans
4. Oliver S. & Ruth Kelly
5. George O.C. & Lucinda Frankenburg
6. Maddox & Mary Fisher
7. Foos Family
8. Charles & Margaret Rabbitts

Section B
9. Pierson & Mary Spinning
10. John & Elmira Ludlow
11. John & Catherine Dick
12. Samuel & Elizabeth Shellabarger

Section C
13. William & Mary Bayley
14. Harry A. & Rosamond Toulmin
15. Henry & Elizabeth Bechtle
16. Robert C. Woodward
17. Samuel A. Bowman
18. General Samson & Minerva Mason
19. Jeremiah & Ann Warder

Section D
20. Jonathan Milhollin
21. Capt. William Addison Stewart
22. James Garfield Stewart
23. John H. & Mary Thomas
24. Daniel & Mary Raffensperger
25. David Lowry

Section E
26. John & Jane Humphreys
27. Ross Mitchell
28. John Crabill
29. James & Mary Ann Leffel
30. Captain Richard Bacon
31. Ezra & Caroline Keller
32. John W. & Rachel Baldwin
33. Col. William & Rachel Werden
34. Charles Turner & Rachel Cavileer
35. Reuben & Mary Miller
36. Snyder Family

Section F
37. George & Sarah Gammon
38. Broadwell Chinn

Section G
39. Max L. Kleeman
40. William N. & Mary Whiteley

Section H
41. Jerome & Martha Uhl
42. Elijah C. & Mary Middleton
43. John & Martha Funk
44. David Orrin Steinberger
45. Dr. Isaac & Clara Kay
46. Col. Charles Anthony
47. General J. Warren Keifer

Section I
48. John S. & Elvira Crowell
49. Sgt. James C. Walker
50. John Ambler & Fanny Shipman
51. Rev. Joseph & Jane Clokey
52. John W. & Eliza Bookwalter

Section K
53. Addison W. & Frances Butt

Section L
54. Robert Layne
55. Dr. Benjamin F. & Ella Prince
56. Rev. Charles & Clara Stroud
57. Hezekiah R. & Nancy Geiger
58. Eliza Daniels "Mother" Stewart
59. Daniel & Catherine Hertzler
60. James & Sarah Fleming
61. Charles A. & Mary Cregar

Section O
62. Gus & Nellie Sun
63. Florence E. & Marietta Kinney
64. Capt. E.L. & Clementine Buchwalter
65. Harry S. & Olive Kissell
66. William H. & Achsa Blee
67. Phineas P. & Anna Mast
68. Merlin G. & Elenore Robertson
69. Dr. Linus Eli & Alice Russell
70. Paul & Elizabeth K. Deer
71. Carleton F. Davidson

Section P
72. Margaret E. Baker
73. Scipio & Jessie Baker
74. Henry L. & Bertha Schaefer
75. Ralph W. & Ellen Hollenbeck
76. Asa S. & Ellen Bushnell
77. William R. & Marcy C. Burnett

Section Q
78. Dr. Bennetta Titlow
79. Newton Hamilton & Lucy Fairbanks
80. McGregor Family
81. William S. & Catherine Gladfelter

Section R
82. Charles F. & Addie McGilvray

Section U
83. Arthur W. & Myrtle Aleshire

Section X
84. Ralph & Edna Davis
85. Harry M. & Violet Turner
86. Michael H. Chakeres

Section Y
87. David "Davey" Moore

Section Z
88. Sully Jaymes

Section 28
89. Bradley Kincaid

Section 32
90. Harry T. & Eugenia Lagos

Mausoleum
91. Robert C. Henry

WWII
92. Richard O. Brinkman

WWII Annex
93. Brooks Lawrence

Ferncliff Cemetery

Biography Index

Aleshire, Arthur W. & Myrtle, 209
Anthony, Colonel Charles & Olive, 173
Bacon, Capt. Richard & Anner, 130
Baker, Margaret E., 222
Baker, Scipio & Jessie, 194
Baldwin, John W. & Rachel, 155
Bayley, William & Mary, 204
Bechtle, Henry & Elizabeth, 132
Blee, William H. & Achsa, 188
Bookwalter, John W. & Eliza, 185
Bowman, Samuel A., 162
Brinkman, Richard O., 223
Buchwalter, Capt. E.L. & Clementine, 183
Burnett, William R. & Marcy C., 200
Bushnell, Asa S. & Ellen, 174
Butt, Addison W. & Frances, 166
Cavileer, Charles Turner & Rachel, 137
Chakeres, Michael H., 228
Chinn, Broadwell, 161
Clemans, George W., 143
Clokey, Rev. Joseph & Jane, 158
Crabill, John & Barbara, 201
Cregar, Charles A. & Mary, 164
Crowell, John S. & Elvira, 193
Davidson, Carleton F., 226
Davis, Ralph & Edna, 218
Deer, Paul & Elizabeth K., 220
Dick, John & Catherine, 178
Fairbanks, Newton H. & Lucy, 205
Fisher, Maddox & Mary, 131
Fleming, James & Sarah, 169
Foos Family, 141
Frankenberg, George O.C. & Lucinda, 177
Funk, John & Martha, 160
Gammon, George & Sarah, 176
Geiger, Hezekiah R. & Nancy, 167
Gladfelter, William S. & Catherine, 179
Henry, Robert C., 219
Hertzler, Daniel & Catherine, 146
Hollenbeck, Ralph W. & Ellen, 206
Hook, Jonathon & Magdalena, 150
Humphreys, John & Jane, 140
Jaymes, Sully, 215
Kay, Dr. Isaac & Clara, 191

Keifer, Gen. J. Warren, 202
Keller, Ezra & Caroline, 133
Kelly, Oliver S. & Ruth, 175
Kincaid, Bradley, 224
Kinney, Florence E. & Marietta, 199
Kissell & Harry S. & Olive, 212
Kleeman, Max L., 208
Lagos, Harry T. & Eugenia, 229
Lawrence, Brooks, 227
Layne, Robert, 133
Leffel, James & Mary Ann, 145
Lowry, David, 143
Ludlow, John & Elmira, 157
Mason,, Gen. Samson & Minerva, 147
Mast, Phineas P. & Anna, 166
McGilvray, Charles F. & Addie, 195
McGregor Family, 189
Middleton, Elijah C. & Mary, 156
Milhollin, Jonathan, 130
Miller, Reuben & Mary, 153
Mitchell, Ross & Catherine, 184
Moore, Davey, 217
Prince, Dr. Benjamin F. & Ella, 203
Rabbitts, Charles & Margaret, 169
Raffensperger, Daniel & Mary, 152
Robertson, Merlin G. & Elenore, 221
Rodgers, Dr. Robert S. & Effie, 154
Russell, Dr. Linus E. & Alice, 188
Schaefer, Henry L. & Bertha, 197
Shellabarger, Samuel & Elizabeth, 163
Shipman, John Ambler & Fanny, 172
Snyder Family, 148
Spinning, Pierson & Mary, 138
Steinberger, David Orrin, 211
Stewart, Eliza Daniels "Mother," 180
Stewart, James Garfield, 215
Stewart, Capt. William Addison, 149
Sun, Gus & Nellie, 216
Thomas, Hon. John H. & Mary, 171
Titlow, Dr. Bennetta, 214
Toulmin, Harry A. & Rosamond, 210
Turner, Harry M. & Violet, 226
Uhl, Jerome & Martha, 187
Walker, Sgt. James C., 196
Warder, Jeremiah & Ann, 135
Werden, Col. William & Rachel, 149
Whiteley, William N. & Mary, 181
Woodward, Robert C. & Lizzie, 165

Bibliography

Bauer, Fern I. *The Madonna of the Trail.* Springfield, 1984.

Beers, William H. *History of Clark County, Illustrated.* Chicago, 1881.

Boyd, Scott Lee. *The Parish Family, Including the Allied Families of Belt-Boyd-Cole and Malone-Clokey-Garrett-Merryman-Parsons-Price-Tipton.* 1935.

Biographical Record of Clark County, Ohio. J.S. Clarke Publishing Co., Chicago & New York, 1902.

Broadstone, Hon. A. *History of Greene County, Ohio: Its History, Industries and Institutions.* B. F. Brown Publishing Co., 1918.

Chapman Brothers. *Portrait and Biographical Album of Greene and Clark County.* 1890.

Clark County, Ohio Probate Records. Clark County Historical Society.

Clippinger, Howard. *Charles Anthony Grand Master.* 1832.

Diehl, Michael. *Biography of Ezra Keller: Founder and First President of Wittenberg College, Springfield, O.* Ruralist Publishing Co., 1859.

Directory of City of Springfield. Stephenson & Co., 1852. Reprint, Clark County Historical Society, 1969.

Dirr, Michael A. *Manual of Woody Landscape Plants.* Fifth Edition. Stipes Publishing, L.L.C., Champaign, Illinois, 1998.

Egle, William Henry, M.D. M.A., Ed. *Notes and Queries, Historical & Genealogical Chiefly Related to Interior Pennsylvania.* Harrisburg, Pennsylvania Publishing Co., 1895.

Early History of the United Presbyterian Church, 1929. Clark County Historical Society.

Ferncliff Cemetery Interment Records; Tombstone Readings.

GFWC/Ohio Federation of Women's Clubs. 2000-2006. *Dedication Held for Statue of OFWC Founder, Clementine Berry Buchwalter.*

Howe, Henry. *Historical Collections of Ohio: General and Local History of Counties, Cities and Villages.* Published by the State of Ohio, Cincinnati, 1904.

Illustrated Historical Atlas of Clark County, Ohio. L.H. Everts and Company, Philadelphia, 1875. Reprint, The Bookmark, Knightstown, Indiana, 1974.

Jackson, Kenneth T. and Camilo Jose Vergara. *Silent Cities: The Evolution of the American Cemetery.* Princeton Architectural Press, 1st edition, 1989.

Keister, Douglas. *Stories In Stone: A Field Guide to Cemetery Symbolism and Iconography.* Gibbs Smith, 2004. Special Edition by MJF Books, New York.

Kinnison, William A. *Samuel Shellabarger (1817-1896): Lawyer, Jurist, Legislator.* Springfield, Clark County Historical Society, 1966.

Kumerow, Burton K. *Heartland.* Annapolis, MD, 2001.

Lake, D.J., C.E. *Atlas of Clark County, Ohio.* Philadelphia, C.O. Titus, 1870.

Littell, John. *Family Records or Genealogies of the First Settlers of Passaic Valley, New Jersey and Vicinity.* Feltville, New Jersey, 1852. Reprint, Genealogical Publishing Co., Baltimore, 1976.

Ludlow, John. *The Early Settlement of Springfield, Ohio.* Springfield, Clark County Historical Society.

Middleton, E.C. President Lincoln Letters. Clark County Historical Society.

Miller, Guy G. *New Boston: Clark County's Vanished Town.* Springfield, Clark County Historical Society, 1955.

Miller, Mary McGregor. *The Warder Family: A Short History.* Springfield, Clark County Historical Society, 1957.

Mitchell, Alan. *The Trees of North America. Facts on File.* New York, NY, 1987.

Moler, J. Douglas. *Maps of Clark County, Ohio*. From Surveys by T. Kizer, 1855, Cincinnati, Middleton, Strowbridge & Co., 1959.

National Archives and Records Administration. Military Service Records. Washington, D.C.

National Society Daughters of the American Revolution. *D.A.R. Patriot Index*. Washington, 1966.

Official Roster I and II: Soldiers of the American Revolution Who Lived in the State of Ohio. The Ohio Society D.A.R., 1929, 1938.

Official Roster III: Soldiers of the American Revolution Who Lived in the State of Ohio. The Ohio Society D.A.R., 1959.

Official Roster of Ohio Soldiers, Sailors, and Marines in World War I, 1917-1918. The J.F. Heer Publishing Co., Columbus, Ohio.

Ohio Soldiers' and Sailors' Orphanage Records. Xenia, Ohio.

Parsons, D.J., M.D. *History of the Medical Society of Clark County, 1815-1955*.

Personal Interviews. George Berkhofer, Grant Edwards, Jerry Feinstein, Eric Mounts, Tom Hopewell, Dan Powell, Marv Wiseman, Stan Spitler, Rod Wyse, Anne Benston, Beth Anderson, Kim Spitler. et. al., 2005-2007.

Prince, Benjamin F. *Standard History of Springfield and Clark County, Vol. I, II*. Chicago and New York, American Historical Society, Inc., 1922

Prince, Benjamin F. *Centennial Celebration of Springfield, Ohio Held August 4th to 10th, 1901*. Springfield Publishing Co.

Raup, George B. *Ross Mitchell: Industrialist, Farmer, 1724-1813*. Clark County Historical Society.

Reeder, Albert. *Sketches of South Charleston, Ohio*. South Charleston, 1910.

Roberds, Calvin E. *150 Years: From Buckets to Diesels, The Springfield Fire Department*. Springfield, Miller Printing Co., 1978.

Rockel, William A., ed. *20th Century History of Springfield and Clark County, and Representative Citizens*. Chicago, Biographical Publishing Co., 1908.

Roster of Ohio Soldiers in the War of 1812. Published under the Authority of Law by the Adjutant General of Ohio, 1916. Reprint, The Genealogical Publishing Co., Baltimore, 1966.

Sesquicentennial of the First United Presbyterian Church, 1817-1967. Clark County Historical Society.

Sketches of Springfield in 1956. Reprint, Clark County Historical Society, 1973.

Skardon, Mary, ed. *Soldiers of the Revolution in Clark County, Ohio, Part I*. Springfield, Clark County Historical Society, 1976.

Skinner, Herbert K. *A Scrapbook of Newspaper Articles on Springfield History*. Springfield History Committee, 1971.

Slager, A.L., ed. *Revolutionary War Soldiers Buried in Clark County*. Columbus, Ohio Archaeological Historical Society Publications, 1929.

Snodgrass, Anne, ed. *Soldiers of the Revolution in Clark County, Ohio, Part II*. Springfield, Clark County Historical Society, 1982.

Springfield Newspapers - *Daily News, Daily Times, Tribune, Press Republic, Republican Gazette, Ohio Pioneer and Clark County Advertiser, Weekly Republic, Western Pioneer, & Sun*.

Tayor's Guide to Trees. Houghton Mifflin Company, Boston, Massachusetts, 1961.

Turner, Edith Ide. *It Happened in Springfield*. The Springfield Tribune Printing Co., 1958.

United States Federal Census: Clark County, Ohio. 1820-1890, 1900-1930.

Urbana, Ohio Newspapers - *Daily Citizen, Citizen & Gazette*.

Woodward, R.G. *Sketches of Springfield, Containing an Account of the Early Settlement; Together With an Outline of the Progress and Improvements of this City Down to the Present Time By a Citizen*. Springfield, T.A. Wick & Co., 1852.

Yesteryear In Clark County. Vols. 1-6. Newspaper interviews with aged citizens in 1893. Reprint, Clark County Historical Society, 1947-1952. Bauer Press.

Images Sources

Benston, Anne E.: Page 150.

Bowman, Samuel A.: Page 162.

Brinkman, Marguerite: Page 223.

Cheek, Richard. Courtesy of Mount Auburn Cemetery, Cambridge, MA.: Page 1.

Clark County Historical Society: Pages 8, 9, 14, 33 (John Ludlow), 131, 133, 135, 136, 138, 142, 143, 145, 147, 148, 149, 155, 156, 160, 163, 166, 169, 171, 174, 175, 180, 181, 183, 184, 185, 187, 188, 189, 191, 194, 195, 196, 199, 200, 201, 202, 203, 205, 206, 208, 210, 211, 214, 215, 216, 217, 222, 224, 226 & 227.

Dallenbach, Tamara K.: Pages 10-11, 46-47, 51, 54-55, 57, 58-59, 69, 74-75, 76-77, 82-83, 90-91, 94-95, 105, 107, 116-117, & cover image.

Deer & Kuss Families: Page 220.

Ferncliff Cemetery: Pages 27, 30, 38, 39, 60, 61, 96, & 97.

Hatfield, Roderick J.: Pages 43, 80, & 99.

John Turner Landess Collection: Pages 45, 140, 146, 193, & 226.

Lagos, James H.: Page 229.

Mount Auburn Cemetery, Cambridge, MA.: Pages 2, 3, 4, & 5.

Noonan, Mary Lu K.: Page 212.

Robert C. Henry Family: Page 219.

Rose, Kevin R.: Pages 20, 23, 25, 37, 40-41, 72-73, 93, & 102-103.

Schahner, Paul W.: Pages 7, 16-17, 24, 34-35, 49, 63, 64-65, 93, 101, 108-115, & 129.

Springfield News-Sun: Pages 218, 228.

Wait, Richard L.: Pages 31, 33 (Ludlow Monument), 66, & 87.

William Bayley Family: Page 204.

Wittenberg University Archives: Page 167.

About the Authors

Anne Benston is a member of the fourth generation of her family to call Clark County home. She graduated from Catholic Central High School and attended Antioch College. Her avid interest in the people of early nineteenth-century Clark County took her to the doors of the Clark County Historical Society in 1970, where she served as a volunteer for many years. Her projects included locating early burial sites and reading and recording headstone inscriptions. She also abstracted historical material from every Clark County newspaper from 1829 to 1842. Her research on Revolutionary War soldiers culminated in the publication of *Soldiers of the Revolution in Clark County, Ohio, (part 11)*, published by the historical society.

Anne served on the board of the historical society board from 1975-1981, and held the offices of treasurer and president. She is a charter member and past president of the Clark County Genealogical Society, and served on the Hertzler House committee in preparation for 1976 bicentennial celebration.

In 1981, the family moved to the central valley of California where she continued her interest in history, serving as treasurer of the McHenry Museum. On returning to Ohio she resumed her volunteer work at the Clark County Historical Society. She also serves as a trustee of the Heritage Commission of South Charleston, as a member of the Ferncliff Cemetery Tour Committee, and as a member of the Springfield Sculpture Advisory Committee.

At the 106th Annual Meeting of the Clark County Historical Society in 2003, the Board of Trustees honored Anne Benston with its highest tribute, the *Dr. Benjamin F. Prince Award*, for her decades of outstanding service.

Anne is married to Charles Benston, and she has two children, Robert and Michelle.

Paul Schanher was born and raised in Springfield, Ohio. He graduated from Springfield North High School, Indiana University, and The Ohio State University College of Dentistry. He is a member of American Dental Association, Ohio Dental Association, and is past president of the Mad River Dental Society. He is in practice in Springfield.

Passionate about local history, he has served on the education committee of the Springfield Preservation Alliance, the Springfield Sculpture Advisory Committee, and as a museum docent at the Heritage Center of Clark County. He regularly speaks to groups on early Springfield and Clark County history. He is also an ardent student of Civil War history.

Paul serves as a member of the Springfield Cemetery Association and conducts walking tours on the history of Ferncliff Cemetery.

He is a member of Southgate Baptist Church and serves as a youth mentor as part of the church's outreach ministry.

Paul and his wife Cheryl have two daughters, Stacy and Stephanie, and a son, Seth.